Diesel Engine Operations with Alternative Fuels

Diesel Engine Operations with Alternative Fuels presents the results of a study that analyzes selected parameters of the combustion process in a diesel engine when fueled with alternative fuels. It discusses the use of a unique test stand consisting of a motor vehicle with a diesel engine adapted to run on different fuels: liquid and gas.

Intending to carry out the process of indicating the engine and measuring exhaust gas toxicity under near real conditions, the book demonstrates the implementation of a Worldwide harmonized Light vehicles Test Procedure (WLTP test). It shares research that seeks alternative fuels to power the diesel engine, including diesel and hydrogen, rapeseed oil with n-hexane and hydrogen, rapeseed oil with n-hexane, rapeseed oil and propane-butane gas, and rapeseed oil with n-hexane and propane-butane gas.

The book will interest academic researchers and graduate students studying alternative fuels, vehicle operations, and engine operations.

Diesel Engine Operations with Alternative Fuels

Rafał Longwic

CRC Press is an imprint of the
Taylor & Francis Group, an **informa** business

First edition published 2025
by CRC Press
2385 NW Executive Center Drive, Suite 320, Boca Raton FL 33431

and by CRC Press
4 Park Square, Milton Park, Abingdon, Oxon, OX14 4RN

CRC Press is an imprint of Taylor & Francis Group, LLC

ISBN: 978-1-032-73855-0 (hbk)
ISBN: 978-1-032-73856-7 (pbk)
ISBN: 978-1-003-46629-1 (ebk)

DOI: 10.1201/9781003466291

Typeset in Times
by Apex CoVantage, LLC

Contents

Preface

The monograph presents the author's results of studies of selected combustion process parameters for a diesel engine when fueled with alternative fuels. The fuel variants considered are diesel and hydrogen, rapeseed oil with n-hexane and hydrogen, rapeseed oil with n-hexane, rapeseed oil and propane-butane gas, and rapeseed oil with n-hexane and propane-butane gas.

The research used a unique test stand consisting of a motor vehicle with a diesel engine adapted to run on different fuels—liquid and gas. The stand also included a two-axle chassis dynamometer, an engine indicating system from AVL, an exhaust gas toxicity measurement system from MACHA, and an engine diagnostic parameter monitoring system from BOSCH. The idea of the research was to carry out the process of indicating the engine and measuring exhaust gas toxicity under near-real conditions. For this purpose, the WLTP test was implemented.

The research is aimed at seeking alternative fuels to power the diesel engine, which still dominates vehicle operation. Such activities can contribute to extending the service life of diesel engines already in use. Indeed, the diesel engine may be phased out as a result of new and stricter exhaust gas toxicity standards. The book provides information on the suitability of various alternative fuels for powering a diesel engine. This information can be used in the operation of these engines. The diesel engine is still the primary source of energy in automobiles, agricultural machinery, and work machines. Powering the diesel engine with alternative fuels can benefit the environment and help diversify the fuel market.

I would like to express my gratitude to several people whose help was significant. To Professor Zbigniew Pater, Rector of Lublin University of Technology, who supported my research both with good advice and financially, and to my research team, who helped me technically in conducting the research. I also thank my family for their support during the preparation of this book.

Author Biography

Rafał Longwic is Professor at the Mechanical Faculty of Lublin University of Technology (Lublin, Poland) and Head of the Department of Motor Vehicles. He graduated with a master's degree in mechanical engineering and motor vehicles. He received his doctoral degree in alternative fuels. Prof. Longwic's main interests are the operation of internal combustion engines under dynamic conditions and alternative fuels. Prof. Longwic is the author of 100 papers in journals and conference proceedings and 8 books.

Introduction 1

According to various estimates, there are around 1.2 billion cars in use worldwide, and this figure is expected to rise to 2 billion in the near future. During the pandemic, due to production difficulties, car sales declined slightly but are now on an upward trend. According to the International Organisation of Motor Vehicle Manufacturers, about 93 million motor vehicles will be sold in 2023, including about 65 million passenger cars and about 28 million commercial vehicles. According to the European Automobile Manufacturers Association (ACEA), among commercial cars, about 97% of vehicles were equipped with diesel engines. Among passenger cars, by October 2023, sales of cars with different propulsion systems were as follows:

- petrol cars—35.9%
- hybrid cars—25.5%
- electric cars—14%
- cars with diesel engines—13.9%
- plug-in hybrids—7.6%
- other—3%

There are two trends in car sales:

- sales of hybrid, electric, and plug-in hybrid vehicles are on the rise—the total sales were close to that of internal combustion engine cars in 2023;
- in the luxury car segment, sales of cars with diesel engines have increased—leading brands, for example, Mercedes, BMW, and Audi, are continually improving their diesel engine design to meet new emissions standards.

It appears that petrol engine and hybrid propulsion will dominate in popular passenger car brands—this technology is cheaper than the diesel engine. In summary, the diesel engine is a significant source of motor vehicle propulsion, particularly for commercial vehicles. Globally, there are a huge number

DOI: 10.1201/9781003466291-1

of diesel-powered automotive vehicles in operation. These engines are also an as yet irreplaceable source of mechanical energy in working machinery—agricultural tractors, agricultural machinery, and construction machinery.

One of the main factors driving the search for new power sources for motor vehicles is the growing awareness of climate change and the need to protect the environment. Traditional internal combustion engines, especially those powered by petrol and diesel, emit certain amounts of carbon dioxide (CO_2) and other harmful substances such as nitrogen oxides (NO_x) and particulate matter. These emissions contribute to global warming and lead to the deterioration of air quality, which has a negative impact on public health. Alternative technologies such as electric vehicles (EVs), plug-in hybrids (PHEVs), and hydrogen propulsion offer significant reductions in emissions, which is key to achieving global climate goals such as those set out in the Paris Agreement. Dependence on fossil fuels such as oil poses a significant challenge to energy sustainability. Reducing this dependence by developing alternative sources of propulsion, such as renewable electricity, can improve energy security and reduce the impact of oil price fluctuations on the global economy.

In response to environmental challenges and climate change, many countries are introducing increasingly stringent emissions regulations and fuel standards. These standards are stimulating the development and deployment of alternative powertrains, while forcing the manufacturers to invest in research and development of new technologies. Examples include the European Union's CO_2 reduction initiatives and electric vehicle subsidy programs in the United States and China. At present, there is a very strong emphasis on the use of biofuels, synthetic fuels, or fuels derived from renewable energy sources in order to reduce the market share of fossil fuels as much as possible. The share of fuels from renewable energy sources is regulated by law through documents such as the European Union's RED (Renewable Energy Directive). The first RED I directive, which was adopted in 2009, aimed to achieve at least a 10% share of renewable energy in the transport sector by 2020 and to increase the share of first-generation biofuels such as biodiesel and bioethanol. The RED II directive adopted in December 2018 introduced new, more ambitious targets for energy from renewable fuels and biofuels. The RED II requirements are to increase the share of energy from renewable sources to at least 14% by 2030 and to gradually reduce the share of first-generation biofuels (i.e., those derived from crops that are also used for food production) and to achieve a 3.5% share of advanced renewable fuels (e.g., second-generation fuels produced from waste) by 2030. The current directive, the RED III, is in force since November 19, 2023. According to it, Member States can choose between a target of 14.5% reduction in greenhouse gas intensity using renewable energy sources and a target of 29% share of renewable energy sources in final energy consumption in transport by 2030. It also introduced a 5.5% requirement for the share of advanced biofuels in renewable energy for transport and a 1% requirement for renewable fuels of non-biological origin.

Intensive work is therefore being carried out in the search for new sources of motor vehicle propulsion. The main strands of this work are:

- electric vehicles, hybrid vehicles, and plug-in hybrid vehicles;
- hydrogen vehicles—combustion of hydrogen in internal combustion engines, hydrogen power for fuel cells; and
- new types of fuels—synthetic fuels; first-, second-, and third-generation biofuels; gaseous fuels—compressed natural gas, liquefied natural gas, biomethane, and hydrogen.

Figure 1.1 provides a schematic representation of the breakdown of alternative fuels for powering internal combustion engines.

In view of the facts presented above, the author does not deny the need to seek new sources of mechanical energy for motor vehicles. However, it cannot be prejudged that in the near future, we will not be using motor vehicles with internal combustion engines. Diesel engines appear to be irreplaceable in some areas of transport. Technical advances in the design of diesel engines make it possible to meet current standards for the emissions of toxic exhaust components. At the same time, there are many vehicles in operation equipped

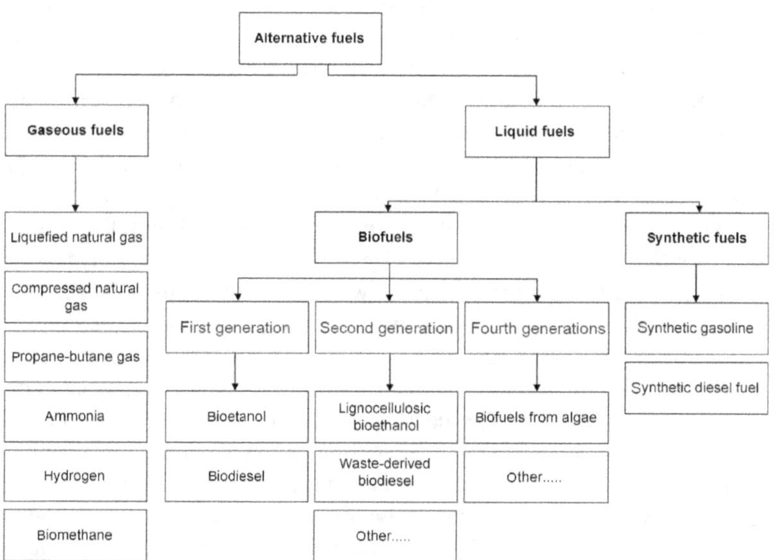

FIGURE 1.1 Breakdown of alternative fuels for internal combustion engines

with internal combustion engines, including diesel engines, which do not meet current standards for emissions of toxic exhaust components. Among other things, in these cases, greener power solutions must be sought. However, these solutions must meet two basic conditions:

- economic condition—the solutions used must be financially acceptable;
- technical prerequisite—the solutions used must be easy to implement, without complex structural changes, and low in energy costs.

The book therefore describes solutions that could be applied to diesel engines currently in operation, meeting the conditions mentioned above. Thus, diesel fuel, rapeseed oil, propane-butane gas, hydrogen, and n-hexane were used in the research presented. Blends of liquid fuels and selected combinations of liquid and gaseous fuels were tested. The fuels tested were therefore:

- diesel with hydrogen,
- rapeseed oil with n-hexane,
- rapeseed oil with n-hexane and hydrogen,
- rapeseed oil with propane-butane gas,
- rapeseed oil with n-hexane and propane-butane gas.

Among other things, the author investigated the possibility of using rapeseed oil in a mixture with n-hexane. This solution is novel. Thanks to the nonreactive solvent, refined rapeseed oil can be used in a diesel engine without esterification. This is a solution that allows the use of rapeseed oil as a fuel to become widespread. It is not necessary to set up industrial rapeseed oil methyl ester production facilities dependent on rapeseed prices and supplies. No energy needs to be expended on the production of rapeseed oil methyl ester. There are no by-products of such production. In addition, the rest of the book presents research into the use of gaseous fuels, including hydrogen, in the diesel engine. The author's intention was to implement the new fuels in the diesel engine as simply as possible, without having to change the diesel fuel dose control maps. The results obtained were related to diesel combustion. This is because diesel is the fuel considered in the construction process of the diesel engine under study. The basic energy indicators of engine operation, the course of combustion, injection, and combustion mixture preparation were analyzed. The tests were conducted under static and dynamic conditions. The entire vehicle was tested on a chassis dynamometer. It was analyzed whether a given fuel type could potentially be used in a diesel engine, taking into account its influence on the course of the combustion process and the emission of toxic components of exhaust gases. The results obtained indicate further possible actions to be taken in order to use a given fuel in operational reality.

Alternative Fuels to Power Diesel Engines

2

2.1 LIQUID FUELS

Chapter 1 mentioned the entire group of liquid alternative fuels. In general, they can be divided into biofuels and synthetic fuels. In the following chapters, the author presented research on the use of first-generation biofuels, mainly rapeseed oil and its blends with a nonreactive solvent. Chapter 2 will discuss the main issues related to the use of rapeseed oil as a fuel and selected scientific papers that treat these issues. The use of liquid alternative fuels in diesel engines, while promising in terms of reducing greenhouse gas emissions and sustainability, faces a number of technological and operational challenges. The most important of these problems are:

1. The physicochemical properties of fuels, mainly those that directly or indirectly (through the fuel injection and atomization process) affect the combustion process—these include viscosity, density, calorific value, surface tension, among others. Rapeseed oil has a higher density (920 kg/m^3) compared to diesel (835 kg/m^3). Rapeseed oil has a significantly higher viscosity (kinematic viscosity index of approximately 37.5 mm^2/s) at 40°C compared to diesel (2.5–4.0 mm^2/s). Rapeseed oil has a significantly higher flash point (around 325°C) compared to diesel (>55°C). Rapeseed oil contains very small amounts of water (0.1%) and sulfur (0 ppm).

DOI: 10.1201/9781003466291-2

Diesel contains 0.02% water and less than 10 ppm sulfur. Diesel has a higher calorific value (42–45 MJ/kg) compared to rapeseed oil (37 MJ/kg). The cetane number of diesel is minimum 51, which is higher than for rapeseed oil (around 50). The fuel's cold filter blocking temperature (CFPP) for diesel is –20 to –32°C, and for rapeseed oil, the CFPP is –5 to –10°C. The surface tension for diesel is about 29 mN/m, which is lower than for rapeseed oil (about 34 mN/m) [1].

2. Chemical composition—variability in the chemical composition of biofuels can affect the combustion process. For example, higher bound oxygen content in biofuels can lead to higher combustion temperatures, which in turn can increase nitrogen oxide (NO_x) emissions. At the same time, it is important to bear in mind that plant feedstocks for fuel production may have different properties depending on the crop [1, 2].
3. Fuel stability and storage—liquid biofuels are more susceptible to degradation during storage. Processes such as oxidation can lead to the formation of deposits that can clog fuel filters and injectors [3–5].
4. Material compatibility—some biofuels can be aggressive toward materials used in conventional diesel engine fuel systems, leading to accelerated wear of components such as gaskets, fuel lines, and pumps [6, 7].
5. Emissions and environmental standards—biofuels can reduce CO_2 emissions, and, at the same time, they can lead to increased emissions of other harmful substances such as nitrogen oxides (NO_x), particulate matter (PM) [8], hydrocarbons (HC), and aldehydes. It is therefore necessary to adapt exhaust gas aftertreatment systems accordingly. The introduction of new fuels is associated with the need to meet stringent exhaust emission standards, which may require costly modifications to existing engines and additional testing and certification [4, 7, 9].
6. Efficiency and operating costs—biofuels differ in calorific value compared to conventional diesel, which affects engine efficiency and fuel consumption. Low energy efficiency can increase operating costs and reduce the economic attractiveness of such fuels. Adapting existing diesel engines to run on biofuels may require modifications to the fuel system, injection system, and combustion chamber, which incur additional costs [4, 9].
7. Availability and infrastructure: the limited infrastructure for the production, distribution, and storage of alternative fuels poses a significant logistical challenge. Investment is needed to expand existing infrastructure to ensure the widespread availability of these fuels [4, 7, 9].

8. Scalability of production: the current scale of production of many biofuels is insufficient to meet the global fuel demand in the transport sector. Developing production technologies and increasing production capacity are key to increasing the availability of these fuels [4, 7, 9].

Biofuels have the potential to significantly reduce the environmental impact of the transport sector, and their widespread use in diesel engines requires overcoming numerous technological, operational, and infrastructural challenges. Further research and innovation in this field are necessary to maximize the potential of these fuels and contribute to the sustainable development of the global transport sector. Plant-based fuels are theoretically characterized by a closed carbon cycle in the atmosphere [9]. When a diesel engine burns biofuel, it emits carbon dioxide, but the plants, which are the source of the fuel, absorb it through photosynthesis and produce oxygen.

Of course, there are alternative liquid fuels that are largely free of the problems caused by their different physicochemical properties to diesel. These include synthetic fuels and, for example, hydrogenated vegetable oil (HVO) [10–12]. Unfortunately, the production process for these fuels is technologically complex and expensive. This translates into the final price of the fuel. In the author's opinion, solutions for liquid alternative fuels that are less costly and less energy-intensive should be sought. The energy intensity of fuel production is an important factor determining its positive impact on the environment. Therefore, the author has devoted much attention to the use of refined rapeseed oil blended with a nonreactive solvent (n-hexane) [13], [14–18]. It should also be borne in mind that the extraction of biofuels from oilseed crops must not be at the expense of food resources. In the case of rapeseed oil, unused land can be used for the cultivation of rapeseed oil.

2.2 GASEOUS FUELS

Gaseous fuels can be used in diesel engines. These include mainly [19–24]:

- compressed natural gas (CNG),
- liquid natural gas (LNG),
- biomethane,
- liquid propane-butane gas (LPG),
- ammonia [25],
- hydrogen.

The gaseous fuels in diesel engines can be used in two ways:

- a design change to the engine to equip it with spark plugs, in which case the diesel engine becomes a spark-ignition engine,
- the combustion of gaseous fuels in the presence of an initiating dose of diesel—so-called dual fuels.

While the use of gaseous fuels in diesel engines offers significant environmental and economic benefits, it also presents a number of technological and operational challenges. Some of these challenges are similar to those listed for liquid fuels. Similar challenges include fuel stability and storage, efficiency and operating costs, availability and infrastructure, and production scalability. Additional challenges include the following:

1. Physical and chemical properties of gaseous fuels, mainly calorific value, density, boiling point, and flash point. The calorific value for propane-butane gas is approximately—for propane 46 MJ/kg and for butane 45 MJ/kg. The calorific value for natural gas is approximately 50–55 MJ/kg. The calorific value for ammonia is approximately 18 MJ/kg. The calorific value for hydrogen is approximately 120 MJ/kg. The boiling point for propane-butane gas is approximately—for propane −420°C and for butane −0.50°C. The boiling point for liquid natural gas is −1,620°C, for biomethane −1,610°C, for ammonia −330°C, and for hydrogen, it is −2,530°C. The density of propane-butane gas is about 2–2.5 kg/m^3, and for natural gas and biomethane, the density is about 0.7 kg/m^3. For ammonia, it is about 0.77 kg/m^3. For hydrogen, it is about 0.09 kg/m^3. The flash point for propane-butane gas is about 470–4950°C for propane and about 365–3,900°C for butane. The flash point for natural gas and biomethane is about 5,400°C. The flash point for ammonia is about 6,300°C, and for hydrogen, it is 5,600°C. The physicochemical properties of gaseous fuels listed above are given under normal conditions.

   In summary, gaseous fuels have a calorific value different than diesel. The heating values of propane-butane gas, natural gas, and biogas are closest to those of diesel fuel. The calorific value of natural gas is subject to strong fluctuations depending on the properties of the raw material used to produce the gas. The calorific value of ammonia is more than twice that of diesel fuel, and the calorific value of hydrogen is more than twice that of diesel fuel. However, the volumetric energy density of gaseous fuels is lower than the volumetric energy density of diesel. This is related to the low density of gaseous fuels. This forces these fuels to be stored under considerable pressure or in the liquid state. The flash point of gaseous fuels is significantly higher than that of diesel.

2. Adaptation of the diesel engine to burn gaseous fuels. Powering a diesel engine with gaseous fuels requires engine design adaptations. The most common form of gaseous fuel combustion in a diesel engine is the combustion of gaseous fuels at the initiating dose of diesel. However, this poses four main problems:
   - the initiating dose of diesel must be controlled accordingly,
   - the gas dosage should be controlled accordingly,
   - an installation should be used which allows the supply and dosage of gaseous fuel—usually sequentially into the intake manifold, and
   - safely store the gas in an additional tank in the vehicle—either in a liquid or gas form.

In this book, in Chapter 4, the author presents a research vehicle that can be fueled with the various gaseous fuels mentioned in the introduction of this chapter with the exception of ammonia [26]. However, there are solutions described in the literature, which allow the engine to be fueled with NH_3 ammonia [25]. However, the combustion of ammonia is associated with a significant problem due to its low calorific value. However, due to its chemical structure, the combustion of both ammonia and hydrogen does not contribute to the formation of carbon dioxide, which is after all a greenhouse gas.

The issue of diesel engine emissions using various gaseous alternative fuels such as LPG, LNG, CNG, biomethane, ammonia, and hydrogen is central to research on reducing the environmental impact of transport [20, 22–24, 26, 27].

Carbon monoxide is a product of incomplete combustion. CO emissions depend on combustion efficiency, which is dependent on the air–fuel ratio (injection control and engine filling) and engine operating conditions. LPG, LNG, CNG, and biomethane typically lead to lower CO emissions as compared to diesel. The above-mentioned gases have fewer carbon atoms in their composition and more hydrogen atoms per carbon atom. Gaseous fuels form combustible mixtures with air, which are better prepared for combustion. Of course, CO emissions when running on LPG, LNG, CNG, and biomethane are affected by the way diesel engines are adapted to burn gaseous fuels, including the gaseous fuel dosage control process.

Both ammonia and hydrogen have no carbon atoms in their composition. The combustion of hydrogen and ammonia should therefore not contribute to the formation of CO. In a diesel engine, however, these fuels do not burn on their own but usually with the initiating dose of diesel fuel, so some CO appears in the exhaust gas, depending on the overall combustion process.

CO_2 emissions are directly related to the carbon content of the fuel. LPG, LNG, CNG, and biomethane can reduce CO_2 emissions per unit of energy compared to diesel due to fewer carbon atoms and better preparation of the combustible mixture with air. Of course, CO_2 emissions are affected by the way diesel engines are adapted to burn gaseous fuels, including the process of controlling the gaseous fuel dose. Ammonia does not contain carbon, so its combustion does not generate CO_2, an important advantage in terms of reducing greenhouse gas emissions. Burning hydrogen also does not generate CO_2, making it an attractive fuel in the context of climate policy.

Incomplete combustion of fuels can lead to HC emissions. Gases such as LPG, LNG, CNG, and biomethane typically have lower HC emissions compared to diesel combustion as a result of better miscibility with air and more efficient combustion. Ammonia and hydrogen do not generate hydrocarbons through combustion as they contain no carbon. Their HC emissions can only be related to possible impurities or additives.

NO_x are produced by high-temperature reactions of nitrogen and oxygen in the air. LPG, LNG, CNG, and biomethane can generate comparable or lower NO_x emissions than diesel, depending on combustion conditions and NO_x reduction technology. However, the possibly higher combustion temperature of gaseous fuels may be an issue here. Ammonia and hydrogen can lead to slightly higher NO_x emissions precisely because of the high combustion temperatures.

There is promise for hydrogen combustion in diesel engines [28–30]. It offers significant benefits in terms of CO_2 emissions reduction and can contribute to the sustainable development of the transport sector. However, the technology faces a number of challenges related to the physicochemical properties of hydrogen, storage safety, NO_x emission control, engine adaptation costs, and lack of infrastructure. To fully realize the potential of hydrogen as a fuel, further research, technology development, and investment in hydrogen production, transport, and storage infrastructure are needed.

REFERENCES

[1] A. Zdziennicka, K. Szymczyk, B. Jańczuk, R. Longwic, and P. Sander, "Surface, volumetric, and wetting properties of oleic, linoleic, and linolenic acids with regards to application of canola oil in diesel engines," Applied Sciences (Switzerland), vol. 9, no. 17, 2019, doi: 10.3390/app9173445.

[2] K. Gorski, R. Smigins, J. Matijošius, A. Rimkus, and R. Longwic, "Physicochemical properties of diethyl ether-sunflower oil blends and their impact on diesel engine emissions," Energies (Basel), vol. 15, no. 11, 2022, doi: 10.3390/en15114133.

[3] V. Vrabie, D. Scarpete, and O. Zbarcea, "Vegetable oils as alternative fuel for new generation of diesel engines: A review," XXIV International Scientific-Technical Conference, vol. 1, 2019.
[4] F. Liu, M. Shafique, and X. Luo, "Literature review on life cycle assessment of transportation alternative fuels," Environmental Technology & Innovation, vol. 32, 2023, doi: 10.1016/j.eti.2023.103343.
[5] B. L. Salvi, K. A. Subramanian, and N. L. Panwar, "Alternative fuels for transportation vehicles: A technical review," Renewable and Sustainable Energy Reviews, vol. 25, 2013, doi: 10.1016/j.rser.2013.04.017.
[6] L. A. Raman, B. Deepanraj, S. Rajakumar, and V. Sivasubramanian, "Experimental investigation on performance, combustion and emission analysis of a direct injection diesel engine fuelled with canola oil biodiesel," Fuel, vol. 246, 2019, doi: 10.1016/j.fuel.2019.02.106.
[7] H. Stančin, H. Mikulčić, X. Wang, and N. Duić, "A review on alternative fuels in future energy system," Renewable and Sustainable Energy Reviews, vol. 128, 2020, doi: 10.1016/j.rser.2020.109927.
[8] W. Lotko, R. Longwic, and M. Swat, "The effect of rape oil—diesel oil mixture composition on particulate matter emission level in diesel engine," SAE Technical Papers, 2001, doi: 10.4271/2001-01-3388.
[9] S. Mathur, H. Waswani, D. Singh, and R. Ranjan, "Alternative fuels for agriculture sustainability: Carbon footprint and economic feasibility," AgriEngineering, vol. 4, 2022, doi: 10.3390/agriengineering4040063.
[10] G. M. Pinto, R. B. R. da Costa, T. A. Z. de Souza, A. J. A. C. Rosa, O. O. Raats, L. F. A. Roque, G. V. Frez, and C. J. R. Coronado, "Experimental investigation of performance and emissions of a CI engine operating with HVO and farnesane in dual-fuel mode with natural gas and biogas," Energy, vol. 277, 2023, doi: 10.1016/j.energy.2023.127648.
[11] A. M. Krais, J. Y. Essig, L. Gren, C. Vogs, E. Assarsson, K. Dierschke, J. Nielsen, B. Strandberg, J. Pagels, K. Broberg, C. H. Lindh, A. Gudmundsson, and A. Wierzbicka, "Biomarkers after controlled inhalation exposure to exhaust from hydrogenated vegetable oil (HVO)," International Journal of Environmental Research and Public Health, vol. 18, no. 12, 2021, doi: 10.3390/ijerph18126492.
[12] L. F. A. Roque, R. B. R. da Costa, T. A. Z. de Souza, C. J. R. Coronado, G. M. Pinto, A. J. A. Cintra, O. O. Raats, B. M. Oliveira, G. V. Frez, and L. F. R. Alves, "Experimental analysis and life cycle assessment of green diesel (HVO) in dual-fuel operation with bioethanol," Journal of Cleaner Production, vol. 389, 2023, doi: 10.1016/j.jclepro.2023.135989.
[13] R. Longwic and J. Kowalczyk, "The influence of the dynamic angle of fuel pumping start on selected parameters of the combustion process in diesel engine powered by mixtures of canola oil with n-hexane," IOP Conference Series: Materials Science and Engineering, vol. 421, 2018, doi: 10.1088/1757-899X/421/4/042049.
[14] R. Longwic and P. Sander, "The course of combustion process under real conditions of work of a traction diesel engine supplied by mixtures of canola oil containing n-hexane," IOP Conference Series: Materials Science and Engineering, vol. 421, 2018, doi: 10.1088/1757-899X/421/4/042050.
[15] K. Górski, P. Sander, and R. Longwic, "The assessment of ecological parameters of diesel engine supplied with mixtures of canola oil with n-hexane," IOP Conference Series: Materials Science and Engineering, vol. 421, 2018, doi: 10.1088/1757-899X/421/4/042025.

[16] P. Sander, R. Longwic, B. Jańczuk, A. Zdziennicka, and K. Szymczyk, "The use of canola oil, n-hexane, and ethanol mixtures in a diesel Engine," SAE International Journal of Fuels and Lubricants, vol. 14, no. 2, 2021, doi: 10.4271/04-14-02-0008.

[17] R. Longwic, P. Sander, and D. Tatarynov, "Ecological aspects of using mixtures of canola oil with n-hexane in diesel engine," Combustion Engines, vol. 190, no. 3, 2022, doi: 10.19206/CE-143245.

[18] R. Longwic, P. Sander, A. Zdziennicka, K. Szymczyk, and B. Jańczuk, "Combustion process of canola oil and n-hexane mixtures in dynamic diesel engine operating conditions," Applied Sciences (Switzerland), vol. 10, no. 1, 2020, doi: 10.3390/app10010080.

[19] Z. Liu, Z. Guo, X. Rao, Y. Xu, C. Sheng, and C. Yuan, "A comprehensive review on the material performance affected by gaseous alternative fuels in internal combustion engines," Engineering Failure Analysis, vol. 139, p. 106507, 2022, doi: 10.1016/j.engfailanal.2022.106507.

[20] M. Staš, J. Kroufek, T. Hlinčík, and P. Šimáček, "Properties and analysis of gaseous alternative fuels II: Fuels based on natural gas and biogas," Paliva, vol. 14, no. 3, 2022, doi: 10.35933/paliva.2022.03.05.

[21] S. H. Hosseini, A. Tsolakis, A. Alagumalai, O. Mahian, S. S. Lam, J. Pan, W. Peng, M. Tabatabaei, and M. Aghbashlo, "Use of hydrogen in dual-fuel diesel engines," Progress in Energy and Combustion Science, vol. 98, p. 101100, 2023, doi: 10.1016/j.pecs.2023.101100.

[22] O. Park, P. S. Veloo, N. Liu, and F. N. Egolfopoulos, "Combustion characteristics of alternative gaseous fuels," Proceedings of the Combustion Institute, vol. 33, no. 1, pp. 887–894, 2011, doi: 10.1016/j.proci.2010.06.116.

[23] V. N. Nguyen, S. K. Nayak, H. S. Le, J. Kowalski, B. Deepanraj, X. Q. Duong, T. H. Truong, D. N. Cao, and P. Q. P. Nguyen, "Performance and emission characteristics of diesel engines running on gaseous fuels in dual-fuel mode," International Journal of Hydrogen Energy, vol. 49, pp. 868–909, 2024, doi: 10.1016/j.ijhydene.2023.09.130.

[24] N. N. Mustafi, R. R. Raine, and S. Verhelst, "Combustion and emissions characteristics of a dual fuel engine operated on alternative gaseous fuels," Fuel, vol. 109, 2013, doi: 10.1016/j.fuel.2013.03.007.

[25] M. Djermouni and A. Ouadha, "Thermodynamic analysis of methanol, ammonia, and hydrogen as alternative fuels in HCCI engines," International Journal of Thermofluids, vol. 19, 2023, doi: 10.1016/j.ijft.2023.100372.

[26] D. Tatarynov, R. Longwic, P. Sander, Ł. Zieliński, M. Trojgo, W. Lotko, and P. Lonkwic, "Test stand for a motor vehicle powered by different fuels," Applied Sciences (Switzerland), vol. 12, no. 20, 2022, doi: 10.3390/app122010683.

[27] M. Andrych-Zalewska, Z. Chlopek, J. Merkisz, and J. Pielecha, "Research on exhaust emissions in dynamic operating states of a combustion engine in a real driving emissions test," Energies (Basel), vol. 14, no. 18, 2021, doi: 10.3390/en14185684.

[28] G. Wozniak, R. Longwic, K. Szydło, A. Kryłowicz, J. Kryłowicz, and R. Juszczak, "The efficiency of the process of coal gasification in the presence of hydrogen," E3S Web of Conferences, 2018, doi: 10.1051/e3sconf/20184600030.

[29] G. Wozniak and R. Longwic, "Initial assessment of the course of combustion process in a diesel engine powered by diesel oil and Brown's gas," IOP Conference Series: Materials Science and Engineering, vol. 421, 2018, doi: 10.1088/1757-899X/421/4/042085.

[30] S. Kryshtopa, K. Górski, R. Longwic, R. Smigins, L. Kryshtopa, and J. Matijošius, "Using hydrogen reactors to improve the diesel engine performance," Energies (Basel), vol. 15, no. 9, 2022, doi: 10.3390/en15093024.

Combustion Process in Diesel Engine

3

3.1 INFLUENCE ON THE COMBUSTION PROCESS OF PHYSICOCHEMICAL PROPERTIES OF FUELS

Chapter 2 discusses the differences in physical and chemical properties of selected alternative fuels. What is clear is that the combustion process is directly influenced by the calorific value of the fuel and its density. Density, together with calorific value, determines the volumetric energy density of fuels. In the current chapter, the author will focus on liquid fuels and the influence of their physicochemical properties on the injection process and the preparation of the combustion mixture in a diesel engine. These processes directly affect the combustion process. One of the main physicochemical parameters of the fuel impacting the injection process, and therefore indirectly the combustion process, is the already mentioned viscosity. The use of a higher viscosity fuel will increase the resistance to movement (friction) of the injection equipment components moving with each other (high-pressure pump components, pressure control valve, atomizer needle). The force of resistance to movement can be determined from relation (3.1).

$$P = \eta \cdot \frac{\pi \cdot d \cdot L}{\delta} \cdot u \tag{3.1}$$

Where:

η—dynamic viscosity of the fuel,

d—diameter of the moving element,

DOI: 10.1201/9781003466291-3

L—length of the lateral surface of the moving element in contact with the fuel in the gap,
δ—the thickness of the fuel layer between the moving element and the body wall, and
u—speed of movement of the moving element.

As the drag force of the movement is directly proportional to the viscosity of the fuel, an increase in this force will result in an increase in the power required to drive the high-pressure pump. At the same time, an increase in needle drag force can result in an increase in needle settling time, which can act to increase injection duration and slightly increase application rate.

Higher fuel viscosity will also increase the hydraulic flow resistance in the injection system. These resistances, expressed in terms of pressure loss in, for example, a circular injection line, can be determined from the Darcy–Weisbach formula (3.2).

$$\Delta p = \lambda \cdot \frac{L \cdot u^2}{2 \cdot d} \cdot \rho \tag{3.2}$$

Where:
λ—friction loss coefficient,
L—the length of the duct section to which the friction loss is related,
u—average flow velocity of the liquid,
d—the internal diameter of the pipe in which the flow takes place, and
ρ—fuel density.

The coefficient of frictional loss can be given in the form of Colebrook and White's empirical formula (3.3).

$$\lambda = \left\{ \frac{1}{-2\log[(k/3,7d) + (6,81/Re)]^{0,9}} \right\}^2 \tag{3.3}$$

Where:
k—the absolute average height of the unevenness of the inner surface of the duct and
Re—Reynolds number.

In order to achieve the required fuel delivery, at the same speed of the high-pressure pump, an additional increase in the power required to drive the pump will therefore be necessary. In addition, with higher fuel viscosity, higher residual pressures in the high-pressure line can be expected. Increasing the fuel viscosity, causing an increase in the resistance to movement and hydraulic flow resistance, will directly affect the movement of the nozzle needle. According to the results of tests on engine feeding with rapeseed oil [1], this should become apparent in an increase in the damping of the needle movement and contribute to the stabilization of nozzle operation.

Less investigated appears to be the effect of viscosity on the values of flow rate coefficients at flow and inlet ports of the pump cylinder and delivery element (effect on element filling) and at the spray ports of the injector. However, based on the above considerations, it is possible to predict an increase in hydraulic flow resistance with an increase in fuel viscosity. This is because the flow through these constrictions will involve local losses, resulting in a reduction in the active flow cross-sections relative to the geometric size. This reduction is usually taken into account by the empirical flow rate coefficient μ, appearing in the Torricelli equation, which determines the fluid discharge q, flowing through the constriction (3.4).

$$q = \mu \cdot f \cdot \sqrt{\frac{2\Delta p}{\rho}} \tag{3.4}$$

Where:
μ—flow rate coefficient;
f—theoretical cross-sectional area of the flow, determined from the geometric dimensions;
Δp—the difference in pressure prevailing on both sides of the constriction;
ρ—fuel density.

The flow rate coefficient can be expressed by the empirical relationship (3.5).

$$\mu = \mu_0 + \frac{\Delta \nu}{3} \tag{3.5}$$

Where:
μ—the coefficient value sought for a fuel with viscosity ν,
μ_0—known value of coefficient at flow of fuel of known viscosity ν_0,
$\Delta\nu = \nu_0 - \nu$, the difference in kinematic viscosity of the two fuels.

Based on the above considerations, it can be concluded that the flow rate coefficient will decrease with increasing viscosity. The active flow cross-section will therefore decrease.

The factors discussed above will cause (at the same rotational speed and constant position of the fuel dose control organ, for a given type of fuel) a worsening of the filling of the working spaces in the high-pressure pump. At the same time, it should be borne in mind that with higher fuel viscosity, there will be a reduction in leakage in the delivery section of the high-pressure pump and the atomizer. This can be justified by Poiseuille's relationship that is [2]:

$$q = \frac{\psi \cdot a^3 \cdot \Delta p}{12 \cdot \eta \cdot L} \tag{3.6}$$

Where:
ψ—slot width,
a—the instantaneous gap height (clearance between the high-pressure pump delivery components),
Δp—pressure differential along the length of the gap,
η—dynamic viscosity of the fuel,
L—slot length.

From the formula given, it can be seen that for given design (ψ, a, L) and hydraulic (Δp) conditions, the amount of leakage is inversely proportional to the viscosity of the fuel.

The physicochemical properties, in particular viscosity, density, and surface tension, will affect the fuel atomization process determined by the parameters of the combustible mixture. Viscosity is an important determinant of the fuel atomization process. This can be seen from the relationship that determines the diameters of the droplets formed after leaving the nozzle outlet (3.7).

$$d_k = \frac{6 \cdot \sigma}{E_p - A \cdot \eta \cdot u} \tag{3.7}$$

Where:
d_k—droplet diameter,
σ—fuel surface tension,
E_p—fuel pulsation energy,
A—fixed,
η—fuel dynamic viscosity index,
u—fuel flow rate.

From the relationship given, it is clear that an increase in viscosity leads to an increase in the fuel droplet diameter. In fact, the fuel droplets undergo further breakdown (secondary breakdown of fuel droplets). The quality of this breakdown is indicated by the mean Sauter diameter, that is, the diameter of a homogeneous surrogate set of droplets with the same total volume and the same total area of all droplets as in a given set with a given spray spectrum. The average Sauter diameter can be calculated on the basis of the empirical formula given by Hiroyasu and Kadota [3]. An increase in the diameter of the fuel droplets does not necessarily lead to an increase in the total jet range. Although droplets with a larger diameter have a higher kinetic energy, their secondary decay occurs earlier. Secondary droplet breakup occurs due to the aerodynamic drag force on the droplet from the injected medium. The droplet undergoes secondary disintegration when the aerodynamic drag force exceeds the surface tension force in the droplet. Larger-diameter droplets have a larger frontal area, and, therefore, their secondary breakup should occur earlier.

In summary, the effect of fuel viscosity on injection and combustible mixture formation is multidirectional. It cannot be stated unequivocally that an increase in fuel viscosity will result in an increase in volumetric dose. In systems with tray diesel injection, this influence will be small although possible. The difference in density of rapeseed oil-based fuels relative to diesel should not significantly affect the formation of the combustible mixture, although fuel density determines many of the parameters of the spray pattern. For the same volumetric doses of fuels with different densities and the same calorific value, the energy dose will be higher for the higher density fuel. This relationship will be directly proportional.

3.2 COURSE OF COMBUSTION PROCESS UNDER STATIC AND DYNAMIC CONDITIONS

The motor vehicle engine mainly operates under variable speed and load conditions. Testing of the internal combustion engine under static conditions (external characteristics, load characteristics) is no longer predominant, mainly for exhaust emission tests [1, 2, 4–8]. Examples are the Worldwide harmonized Light vehicles Test Procedure (WLTP) driving test and real-world vehicle tests. With this in mind, the author has presented both motor vehicle engine tests under static and dynamic conditions in the book.

A dynamic process of engine operation will occur when there is variation in time of the engine speed or the useful engine torque or the total heat flow through the engine [8, 9]. The variation in the heat flux through the engine is most often caused under operating conditions by a change in the amount of energy supplied with the fuel. The usable torque, on the other hand, depends on the engine speed, the amount of fuel supply and the value of the internal loss torque, and the total efficiency of the engine. Analyzing the man–machine–environment system, it is possible to distinguish three groups of causes that can induce a change in engine speed (n), useful engine torque (M_0), and the total heat flow through the engine:

- a change in the load torque on the engine or a change in the internal loss torque,
- change in fuel supply,
- changing the air supply.

The periodicity or non-periodicity of the above changes will affect the occurrence of steady-state or transient engine operating conditions. Due to the interaction between the engine, its controllers (speed, load, ignition angle, etc.), and the human—the dynamic process of engine operation can take place in three variants:

- M_0 = Constant and n ≠ constant,
- M_0 ≠ Constant and n = constant, and
- M_0 ≠ Constant and n ≠ constant.

A full breakdown of engine processes taking into account their periodicity is presented in. This breakdown shows that all engine processes can be classified according to their dynamic properties and the role they play in engine operation in:

- processes associated with individual engine cycles and processes that are rapidly variable in relation to the processes associated with engine traction operation,
- processes associated with engine traction operation,
- processes related to the operational wear of the engine (tribological) and slow-variable processes in relation to the processes related to the traction operation of the engine.

The first group of processes includes all processes related to and conditioning the engine cycle, for example, wave processes in the airflow through the intake system, fuel vapor and fuel vapor diffusion processes, pre-oxidation processes, and fuel combustion. This group can also include processes not always entirely related to the cyclicity of engine operation, although most often generated by it, for example, vibrations of engine components. These processes may have characteristic frequencies much higher than those directly related to the engine operating cycles. In single-engine cycles, rapidly varying processes are decisive for the formation and emission processes of the exhaust components. The characteristic frequencies of these processes must be at least an order of magnitude greater than the frequencies of the engine cycles, as they occur at times corresponding to parts of single cycles, for example, during the rapid increase in pressure of the working medium in the cylinder during combustion. These are therefore frequencies greater than a few tens of hertz.

The second group of processes is related to the operational forces acting on the motor. The primary tractive forcing are the engine control processes and the load from the power receiver. The vehicle response is the speed of the vehicle, and the engine response is the speed of the engine. By the term engine control, the driver's interaction with the engine is meant, that is, the processes caused by the movement of the accelerator lever. The load on the engine is the drag torque (including the moment of inertia of rotating bodies) from the vehicle being driven (resistance to motion and the vehicle's own resistance associated with the drive of equipment such as the alternator, hydraulic power steering pump, and air conditioning compressor). The dynamic properties of the engine control process under traction operation conditions are determined by factors related to the design of the control system and the characteristics of the driver: physiological, volitional, and praxeological. Studies of engine control processes indicate that the shortest time to depress the accelerator pedal is longer than 0.1 s [7]. This means that the characteristic frequencies of this process are certainly less than 10 Hz. At the same time, the longest acceleration times do not exceed several tens of seconds. It can therefore be assumed that the lower limit frequency of the processes characterizing traction operation is greater than 0.01 Hz.

The dynamic properties of the engine load torque are primarily related to the dynamic properties of the vehicle motion, and these in turn depend on the engine control and the resistance to motion. Since inertial components play an important role in the dynamic system "engine control—vehicle—vehicle speed", it can be concluded that the characteristic frequencies of the drag torque process are lower than those of the engine control process. However, at least three additional factors generating the high-frequency drag torque forcing must be taken into account, namely the vibration processes in the drive train and the rapidly varying loads from the interaction of the wheels with the road surface and from the turbulent drag of the body through the air. The amplitudes of these processes are, however, considerably smaller in relation to the amplitude of the basic process; this is primarily due to the existence of damping elements in the power train. It can therefore be assumed that the characteristic frequencies of the processes associated with engine traction operation are less than 10 Hz.

The slow-variable processes in relation to the processes involved in the traction operation of engines are of a completely different dynamic nature to those considered so far. The category of slow-variable processes can include thermal processes that characterize the thermal state of the engine. These processes have time constants of the order of a few minutes or more or at least a few tens of seconds. Processes that are even more slow-variable than thermal processes are those that describe atmospheric conditions and even more so the tribological processes associated with engine wear.

A separate group of dynamic processes is the engine-starting process. The nature of the changes in the parameters of the engine operating processes at start-up depends on many factors, mainly related to the specifics of the engine design, the start-up system used, the temperature of the combustion chamber walls, the type of oil and coolant used, the ambient parameters, etc. [9].

The free acceleration process is used as a method to determine the maximum useful power, or maximum useful torque, of an engine. It is therefore an alternative method to the method of measurement under external operating characteristics of the engine. The following statements are true with regard to the free acceleration process of the engine:

- both during the conditions of external characteristics and during the free acceleration process, the fuel dose control lever is in the maximum position,
- under the conditions of external characteristics, there is a state of equilibrium between the useful engine torque and the external load torque on the engine; under the conditions of free acceleration, there is an instantaneous state of equilibrium between the useful engine torque and the load torque, which is the moment of inertia of the rotating bodies associated with the crankshaft.

The possibility of making comparisons between the combustion parameters under free acceleration conditions and those on the external operating curve is further determined by the equality of the injected fuel dose. Due to the same position of the control unit for the injected fuel dose, it should theoretically be similar under the conditions of free acceleration and external characteristics.

Some differences can be caused mainly by:

- an increase of residual pressures in the injection system under free acceleration conditions compared to the external characteristic conditions—under free acceleration conditions, we have a short duration of individual engine cycles necessary to relieve the high-pressure lines after the injection process (the engine speed increases in successive cycles),
- different values of the counter-pressure prevailing in the combustion chamber during the fuel injection period for free acceleration and external characteristic conditions,

- different start and end angles of injection and, as a result, different injection duration angles—this is due to the fuel dosage control process,
- differences in injection pressure waveforms, including maximum injection pressure, and
- differences in the speed values at which the injection process takes place.

Precise measurement of the fuel dose injected during the free acceleration process is very difficult to achieve. A simpler method is to compare parameters describing the injection process that have a direct effect on the dose. These include injection start angle, injection end angle, and maximum injection pressure. For engines with a classic injection system (working machines, agricultural machines), it was observed that the value of the end-of-injection angle under the conditions of free acceleration and external characteristics was similar. A similar phenomenon occurred with regard to the injection start angle. The values of this angle were slightly higher for the acceleration conditions. Under static conditions, successive engine cycles occur at a constant value of average engine speed. Under free acceleration conditions, each successive engine cycle occurs at a higher engine speed.

In the process mentioned in this book, the free acceleration process of the engine was not used as a test method. The vehicle was accelerated in direct gear. The intensity of this acceleration was significantly lower than that of the free engine acceleration process. Since the free acceleration process is characterized by the highest intensity (measured by the maximum angular accelerations of the engine crankshaft), a discussion of the specifics of this process was important from the point of view of dynamic engine operating conditions. This is because the phenomena observed during free acceleration of the engine will also occur in acceleration processes with lower intensities.

REFERENCES

[1] R. Longwic, P. Sander, A. Zdziennicka, K. Szymczyk, and B. Jańczuk, "Combustion process of canola oil and n-hexane mixtures in dynamic diesel engine operating conditions," Applied Sciences (Switzerland), vol. 10, no. 1, 2020, doi: 10.3390/app10010080.

[2] R. Longwic and P. Sander, "The characteristics of the combustion process occurring under real operating conditions of traction," IOP Conference Series: Materials Science and Engineering, vol. 148, 2016, doi: 10.1088/1757-899X/148/1/012071.

[3] J. Tian, Y. Liu, F. Li, and K. Han, "Experimental study on spray characteristics of acetanol biodiesel and modification of spray tip penetration model," Physics of Fluids, vol. 33, no. 9, 2021, doi: 10.1063/5.0063572.
[4] D. Chen, S. Zhao, C. Wei, and Y. Chen, "Research and applications of condition monitoring and predictive maintenance of marine diesel engines," Zhongguo Jixie Gongcheng/China Mechanical Engineering, vol. 33, no. 10, 2022, doi: 10.3969/j.issn.1004-132X.2022.10.004.
[5] A. Sapit, M. A. Razali, M. F. Hushim, M. Jaat, A. N. Mohammad, and A. Khalid, "Dynamic behavior of canola oil spray in diesel engine," Applied Mechanics and Materials, vol. 773–774, 2015, doi: 10.4028/www.scientific.net/amm.773-774.520.
[6] C. Sui, P. de Vos, D. Stapersma, K. Visser, H. Hopman, and Y. Ding, "Mean value first principle engine model for predicting dynamic behaviour of two-stroke marine diesel engine in various ship propulsion operations," International Journal of Naval Architecture and Ocean Engineering, vol. 14, 2022, doi: 10.1016/j.ijnaoe.2021.100432.
[7] R. Longwic, "Dynamic aspects of the diesel engine work," SAE Technical Papers, 2007, doi: 10.4271/2007-01-4210.
[8] M. Andrych-Zalewska, Z. Chlopek, J. Merkisz, and J. Pielecha, "Impact of the internal combustion engine thermal state during start-up on the exhaust emissions in the homologation test," Energies (Basel), vol. 16, no. 4, 2023, doi: 10.3390/en16041937.
[9] M. Andrych-Zalewska, Z. Chlopek, J. Merkisz, and J. Pielecha, "Research on exhaust emissions in dynamic operating states of a combustion engine in a real driving emissions test," Energies (Basel), vol. 14, no. 18, 2021, doi: 10.3390/en14185684.

Experimental Studies 4

4.1 FUELS USED FOR THE STUDY AND THEIR PHYSICOCHEMICAL PROPERTIES

Diesel, rapeseed oil, propane-butane gas, hydrogen, and n-hexane were used in the study. Blends of liquid fuels and selected combinations of liquid and gaseous fuels were tested. The fuels tested were thus:

- hydrogen-diesel,
- rapeseed oil with n-hexane,
- rapeseed oil with n-hexane and hydrogen,
- rapeseed oil with propane-butane gas,
- rapeseed oil with n-hexane and propane-butane gas.

Of course, the reference fuel was diesel, to which other fuels were compared.

The basic physicochemical properties of the fuels tested are listed in Table 4.1:

 DOI: 10.1201/9781003466291-4

TABLE 4.1 Basic physical and chemical properties of the tested fuels

PHYSICOCHEMICAL PROPERTY	*DIESEL*	*RAPESEED OIL*	*N-HEXANE*	*A MIXTURE OF 10% DIESEL FUEL WITH N-HEXANE*	*HYDROGEN*	*PROPANE-BUTANE*
Density at 20°C [kg/m^3]	840	916	660	895.4	–	–
Surface tension at 20°C [mN/m]	29.15	34.15	18.4	30.08	–	–
Kinematic viscosity index at 40°C [mm^2/s]	2.7	34.89	0.5	30.08	–	–
Flash point [°C]	72	199	–22	40	Flammable gas	–95 (propane), –60(butane)
LC cetane number	51.2	53.3	–	73.9	–	–
Calorific value [MJ/kg]	43	36	–	35.8	120	46

4.2 RESEARCH OBJECT

The research was conducted using a test vehicle (Fiat Qubo 1.3 Multijet) equipped with a compression-ignition engine with a common rail system. The basic technical data of the engine is shown in Table 4.2.

4.3 STAND TEST

Traction tests were conducted using a Dynorace DF4FS-HLS chassis dynamometer. The test stand is shown in the block diagram depicted in Figure 4.1.

The test vehicle could be fuelled with diesel fuel from the vehicle's main tank, other mixtures of liquid fuels from an installed auxiliary tank, and gaseous fuels (propane-butane gas, hydrogen). The supply of gaseous fuels was carried out using an installation consisting of gas cylinders, regulators enabling the setting of the supply pressure, hoses, electromagnetic gas injectors located in the intake manifold, and a computer controlling

TABLE 4.2 Basic engine specifications of the test vehicle

Number of cylinders	4
Cylinder diameter [mm]	69.6
Piston stroke [mm]	82
Total volume [cm^3]	1,248
Maximum power [kW]	70
Engine speed at maximum power [rpm]	4,000
Maximum torque [Nm]	200
Speed at maximum torque [rpm]	1,500
Idle speed [rpm]	850 ± 20
Compression ratio	16.8:1
Gear ratio I	3.64:1
Gear ratio II	1.95:1
Gear ratio III	1.28:1
Gear ratio IV	0.98:1
V gear ratio	0.77:1
Main gear ratio	3.56:1

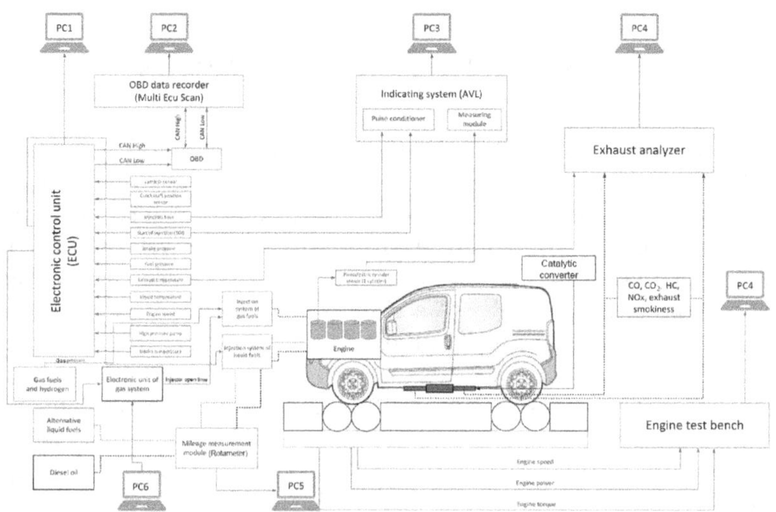

FIGURE 4.1 Block diagram of the test stand

the opening time of the injectors and the moment when the injectors open. The computer controlling the gas system took into account signals from the engine sensors as well as gas temperature and gas pressure. Gas injection was done sequentially into the intake manifold, before each cylinder. Liquid fuel supply was carried out on the basis of the factory's Generation IV common rail diesel fuel system. It was equipped with a set of sensors, actuators (injectors, valves), and a control computer. In the case of supplying liquid fuels other than diesel, an additional fuel tank equipped with a low-pressure pump was used.

The test stand made it possible to record the pressure inside the combustion chamber of the engine, the pressure before the liquid fuel injector, the needle lift of the liquid fuel injectors, engine speed, engine power and torque, and the emission concentration of the toxic component of the exhaust gas. The AVL IndiMicro 602 measurement system with AVL IndiCom software was used to record rapidly varying in-cylinder pressures and injection process parameters. Concentrations of toxic components in the exhaust gas were measured using a MAHA MET 6.3 exhaust gas analyzer, the measuring probe of which was installed upstream of the catalytic converter. The quantities measured by the analyzer were the concentrations of HC, CO, CO_2, O_2, NO_x and λ, as well as the degree of PM opacity. Additional engine operating parameters were recorded using the Multiecuscan OBD II (On Board Diagnostics) diagnostic interface.

4.4 TESTING METHODOLOGY

The testing methodology was based on the idea of the Worldwide harmonized Light vehicles Test Procedure (WLTP). It is a global procedure for testing energy efficiency and emissions for light-duty motor vehicles such as passenger cars. It was introduced to replace the earlier NEDC (New European Driving Cycle) procedure, which was considered inadequate and not reflective of actual fuel consumption and emissions. WLTP features more realistic driving conditions, longer test times, higher speeds, and more representative speed profiles. WLTC (World Harmonized Light Vehicle Test Cycle) is the driving cycle used in the WLTP procedure. It is a set of driving conditions that cover different scenarios such as urban, suburban, highway, and other driving. The WLTC consists of four parts, representing different types of driving, and is used to conduct emissions and fuel consumption tests. Testing motor vehicles in real or near-real conditions is the current trend in determining vehicle properties. So, in the current research, it was decided to use selected points of the WLTP test to evaluate the impact of using new fuels on the performance of a motor vehicle's engine. Since most of the WLTP test runs under dynamic conditions, it was assumed that it is important to test the engine under acceleration conditions. Engine acceleration is an important part of the aforementioned dynamic conditions in the WLTP test. This acceleration occurs at different intensities. The most intensive acceleration occurs under free acceleration conditions, after a sudden surge of the fuel dose control lever. In the conducted research, it was assumed that the tests would take place under dynamic conditions reproduced by accelerating the vehicle in gear II on a chassis dynamometer. The dynamometer's rollers rolled freely. Acceleration was initiated by stepping up the fuel dose control lever. The WLTP test also includes periods of engine operation under static conditions. It was decided to conduct tests at two characteristic points of the WLTP test running under static conditions. The first point was defined by the following parameters: vehicle speed 50 km/h, gear IV, the dynamometer rollers loaded with the value of the moment of forces resulting from the forces of rolling resistance and air resistance acting on the vehicle. The second point was determined by the following parameters: vehicle speed 90 km/h, gear IV, dynamometer rollers loaded with the value of the moment of forces resulting from the forces of rolling resistance and air resistance acting on the vehicle. In addition, the energy suitability of each of the tested fuels was evaluated by a standard vehicle test conducted on a chassis dynamometer. Thus, the tests were conducted under the following conditions:

1. Acceleration test—acceleration of the vehicle in gear II on the chassis dynamometer, the dynamometer rollers rolled freely, acceleration was initiated by stepping up the fuel dose control lever.
2. Static conditions, point I—vehicle speed 50 km/h, gear IV, dynamometer rollers were loaded with the value of the moment of forces resulting from the forces of rolling resistance and air resistance acting on the vehicle.
3. Static conditions, point II—vehicle speed 90 km/h, gear IV, dynamometer rollers were loaded with the value of the moment of forces resulting from the forces of rolling resistance and air resistance acting on the vehicle.
4. A standard vehicle test conducted on a chassis dynamometer to determine maximum engine power and torque.

For condition No. 1, recording of fast-variable parameters of engine operation was carried out, as well as parameters of engine operation parameters were downloaded using the diagnostic interface. For conditions No. 2 and 3, the recording of fast-variable parameters of engine operation was carried out, the parameters of engine operation parameters were downloaded using the diagnostic interface, and the recording of emission concentrations of toxic components of exhaust gases was done. For condition No. 4, the course of power and torque at the wheels of the vehicle were recorded.

Basic Engine Energy Indicators

5

5.1 INTRODUCTION

The book addresses the possibility of using alternative liquid and gaseous fuels in a diesel engine. The main assumption is that this application is possible in the simplest way—without changing the liquid fuel injection control maps. This approach will allow the widespread use of the proposed alternative fuels in vehicle operation. It should be remembered that a diesel engine is designed to burn diesel fuel. The use of alternative fuels changes a number of processes in the engine. These changes affect the values of engine power and torque obtained. This is discussed in the body of the chapter. It should be remembered that maximum engine power and torque are admittedly basic parameters, but they can indicate the economics and environmental impact of the solution. An engine fueled by fuels for which lower values of maximum power and torque were obtained will require higher doses of fuel to overcome certain resistance to motion. These resistances result from the vehicle's resistance to motion. Engine power and torque characteristics were obtained by testing the vehicle on a chassis dynamometer using the direct gear acceleration method. The vehicle's acceleration process represents significant portions of the WLTP test.

5.2 RESEARCH RESULTS

Figure 5.1 shows the power course of the tested engine for three selected fuels—diesel fuel with hydrogen, rapeseed oil with hexane, and rapeseed oil. Figure 5.2, meanwhile, shows the course of engine torque. Figure 5.3 shows

DOI: 10.1201/9781003466291-5

the maximum power values of the tested engine when fed with diesel fuel, diesel fuel with hydrogen, diesel fuel with propane-butane gas, rapeseed oil, rapeseed oil with hydrogen, rapeseed oil with hexane, rapeseed oil with hexane and hydrogen, rapeseed oil with hexane and propane-butane gas and rapeseed oil with propane-butane gas. Figure 5.4 shows the maximum engine torque values when fed with the tested fuels.

The engine was supplied with liquid fuels using a standard fuel path. The engine of the test vehicle, on the other hand, was supplied with gaseous fuels using a dedicated gas system. The gas injectors were located in the engine's intake manifold. Sequential injection of gas into individual cylinders was realized. The gas installation had its own control system, which was coupled to the sensors of the common rail system. Additional injection of gaseous fuel at the liquid fuel supply caused the common rail system controller to limit the dose of liquid fuel based on signals from the engine sensors. The tests presented in

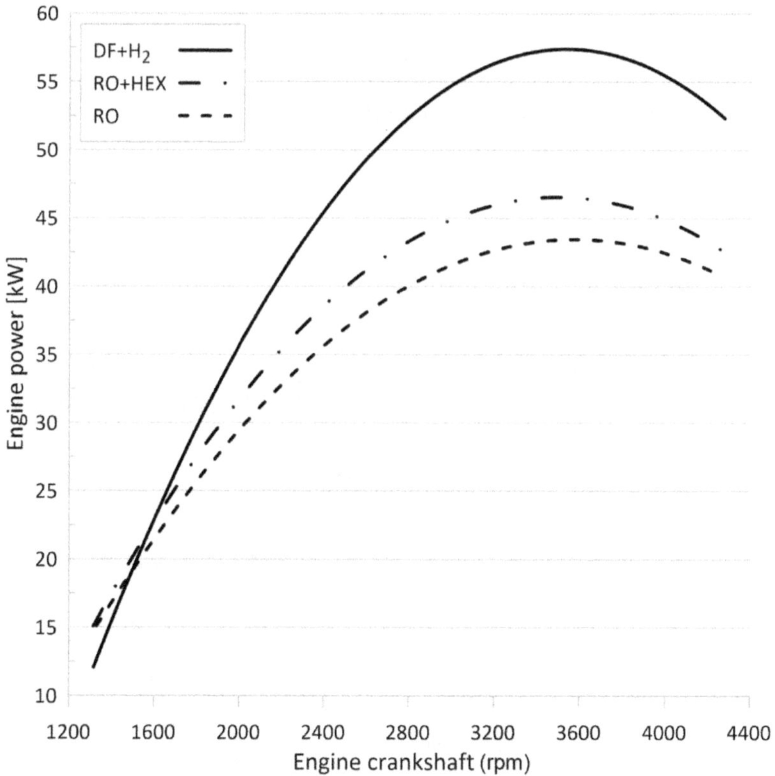

FIGURE 5.1 The course of engine power for selected fuels

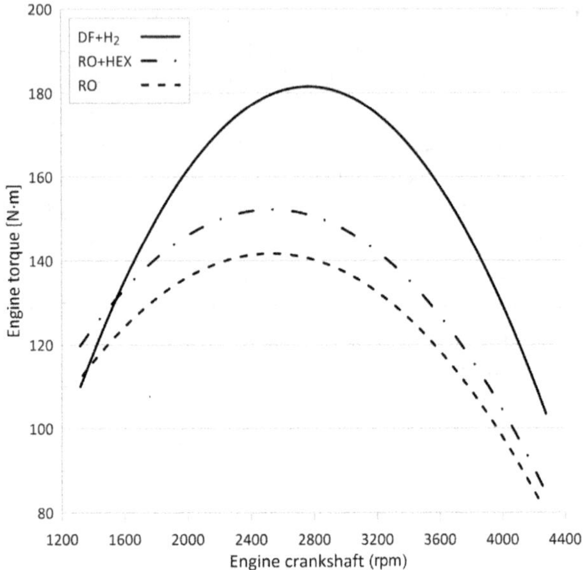

FIGURE 5.2 Engine torque course for selected fuels

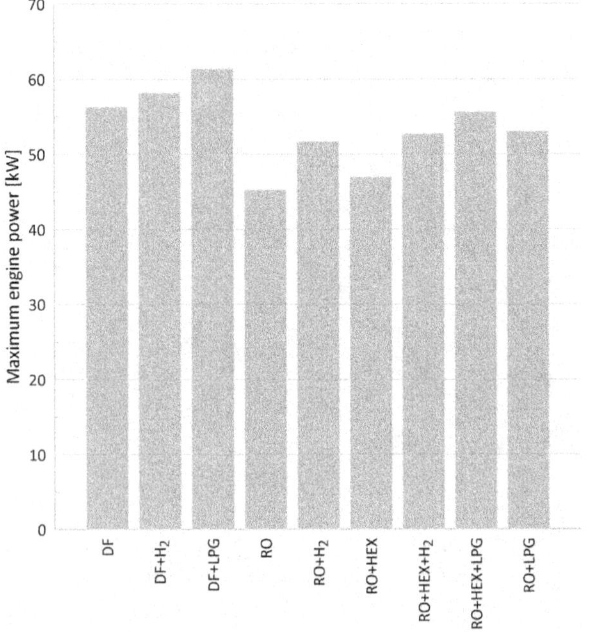

FIGURE 5.3 Maximum engine power values

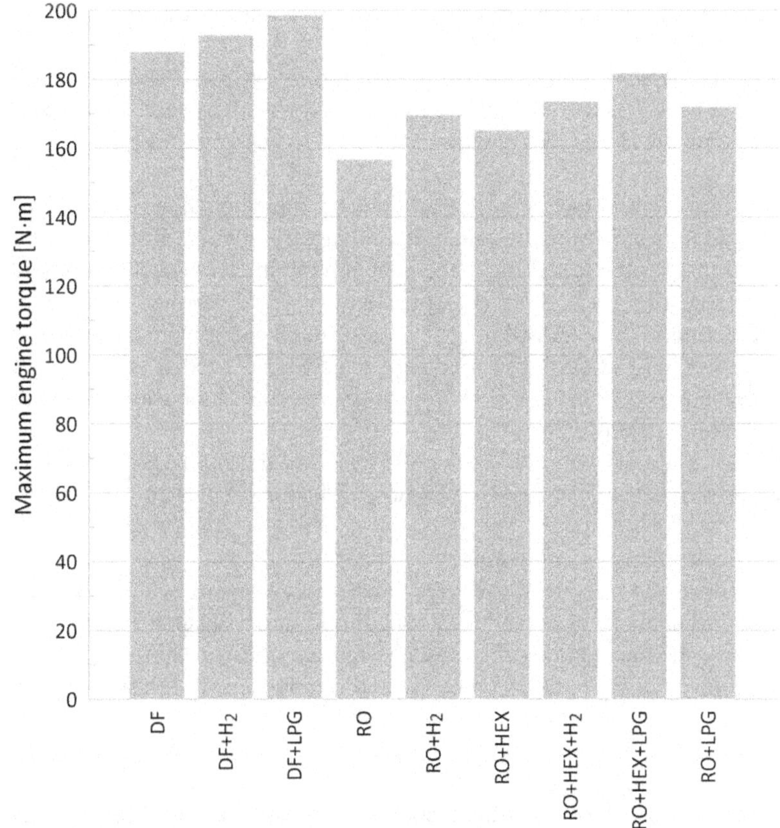

FIGURE 5.4 Maximum engine torque values

this work were carried out for standard settings of the fuel dose control map. This means that the engine's fuel dose control map was not modified. It was the standard map developed by the engine manufacturer for diesel fuel. This is because the idea of the presented research is to be able to use various alternative fuels in a diesel engine in the simplest possible way. The results of the presented research should give knowledge about the necessary changes in the course of the engine control process or exclude the possibility of using certain alternative fuels in operating conditions.

With regard to the engine power and torque obtained with the fuels tested, it should be stated that

- the highest values of maximum engine power and torque were obtained when the engine was fueled with diesel and simultaneously fueled with diesel and propane-butane gas or hydrogen;

- the lowest engine power and torque values were obtained for rapeseed oil;
- increased values of engine power and torque were obtained when hexane was doped into rapeseed oil and when rapeseed oil and gaseous fuels were fed simultaneously.

The maximum engine power and torque values obtained are, of course, the first step in assessing the suitability of the fuels studied. It should be borne in mind that for the tested fuels, the injection parameters were not optimized. For gaseous fuels, a preliminary optimization was performed in this regard. The function of the goal was to achieve maximum energy indicators of engine operation. For LPG, the injector opening time was set to 3 ms and the gas supply pressure to 0.11 MPa. For hydrogen, the injection time was 3 ms, and the supply pressure was 0.25 MPa.

The values of maximum power and torque of the engine when running on alternative fuels are affected by a number of factors, such as:

- calorific value of the fuel;
- mass dose of fuel per engine cycle—in an engine with a common rail injection system, the amount of liquid fuel injected is influenced by the injector opening time and the fuel pressure in the reservoir; the amount of gaseous fuel injected is influenced by the supply pressure and the injector opening time;
- the course of the injection process—the number of pre-injections, the beginning of injection;
- the process of creating a combustible mixture—this process not only depends on the injection pressure and the counter-pressure in the combustion chamber at the time of injection but also very much depends on the physicochemical properties of the fuel, mainly, such as density, viscosity, surface tension, fuel droplet size, secondary disintegration of fuel droplets, the range of the spray pattern of fuel, and the angle of spray of fuel—these issues are discussed in Chapter 3; and
- the amount and swirl of air supplied for combustion.

5.3 CONCLUSION

Based on the tests of engine power and torque when running on alternative fuels, it was found that the highest values of maximum engine power and torque were obtained when running on diesel fuel and simultaneously running on diesel fuel and propane-butane gas or hydrogen. The lowest values of engine power and torque were obtained when running on rapeseed oil. Increased values of engine power and torque were obtained when hexane was doped into rapeseed oil and when rapeseed oil and gas fuels were fed simultaneously. The reasons for the changes in engine power and torque will be explained by analyzing the course of combustion and injection parameters in the next chapter. The explanation of the observed differences is very complex. The alternative fuels used have different chemical and physical properties. They affect a number of processes in the diesel engine. The main ones of these processes are the combustion process, the injection process, and the combustible mixture preparation process.

Selected Parameters of the Combustion and Injection Process

6

6.1 INTRODUCTION

The analysis of the combustion process of the alternative fuels under consideration in a diesel engine concerned the basic parameters of the combustion process (average index pressure, maximum combustion pressure, maximum pressure build-up rate, auto-ignition delay angle), as well as the course of heat release and the amount of heat released. To explain the observed phenomena, the injection process of the studied fuels was also analyzed. The liquid fuel dose for a single injection, the fuel pressure in the common rail system's accumulator and the digital injection signal courses for liquid fuels were analyzed. Explaining the effect of using alternative fuels on the operation of a diesel engine is a very complex issue. The course of the combustion process is influenced by the physical and chemical properties of fuels. They determine the process of preparing the combustible mixture. The physicochemical properties of fuels also affect the injection process. For the different physicochemical properties of alternative fuels compared to diesel, the engine controller selects different injection times and injection start moments. The fuel pressures in the common rail system's accumulator are also different. The above phenomena are tried to be discussed in the following subsections.

DOI: 10.1201/9781003466291-6

6.2 BASIC PARAMETERS OF THE COMBUSTION PROCESS

Figure 6.1 shows the values of the mean index pressure for the tested fuels under conditions corresponding to the movement of a motor vehicle at 50 km/h on a chassis dynamometer. According to subsection 4.4 in Chapter 4, these are conditions similar to operating condition No. 2. This is an excerpt from the WLTP test running under static conditions. The rollers of the chassis dynamometer were loaded with forces equal to the sum of the rolling resistance force and air resistance occurring when the test vehicle moved at 50 km/h on a level road surface. To limit the amount of information, the results of the tests at 90 km/h (Condition No. 3) are not presented. The nature of the course of the analyzed parameters at the vehicle speeds of 50 km/h and 90 km/h was very similar. Figure 6.2 shows the values of maximum combustion pressure, and Figure 6.3 shows the values of maximum pressure rise.

It was found that the values of the mean indicated pressure for diesel fuel were about 0.6 MPa. For cases of simultaneous combustion of diesel fuel

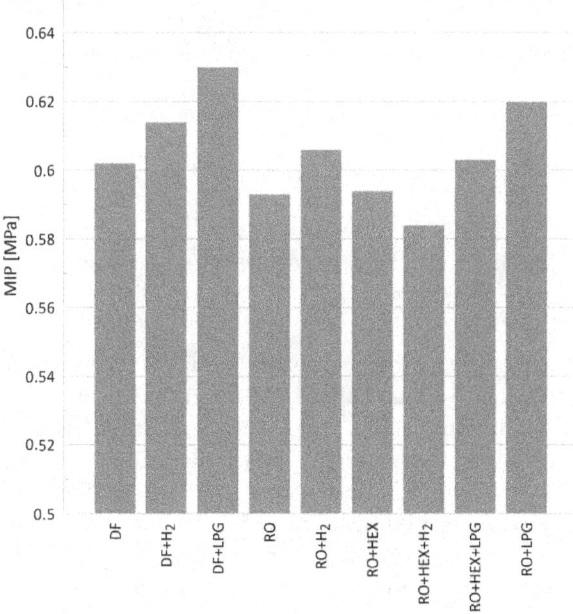

FIGURE 6.1 Mean indicated pressure (MIP), with static conditions, vehicle speed 50 km/h, and gear IV

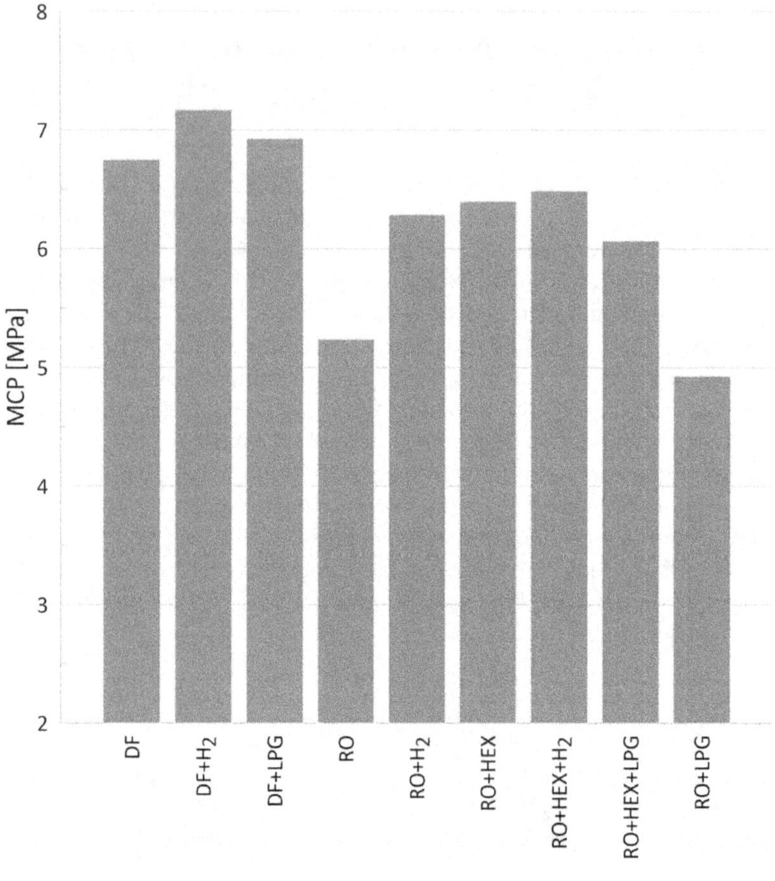

FIGURE 6.2 Maximum combustion pressure MCP, with static conditions, vehicle speed 50 km/h, and gear IV

with hydrogen and propane-butane gas, the values of the mean indicated pressure were higher and were about 0.61 MPa and 0.63 MPa, respectively. The total energy dose of liquid and gaseous fuel was not determined. The method of feeding the engine with gaseous fuels is described in Section 6.2. The smallest values of the mean indicated pressure were obtained for rapeseed oil (about 0.59 MPa) and a mixture of rapeseed oil with hexane and hydrogen. The addition of hexane was mainly aimed at reducing the viscosity of the rapeseed oil and improving the preparation of the combustible mixture and injection. For mixtures of rapeseed oil with hexane, the desired effect was obtained, and the values of mean indicated pressure increased slightly, compared to the use of rapeseed oil only. The case of simultaneous combustion of

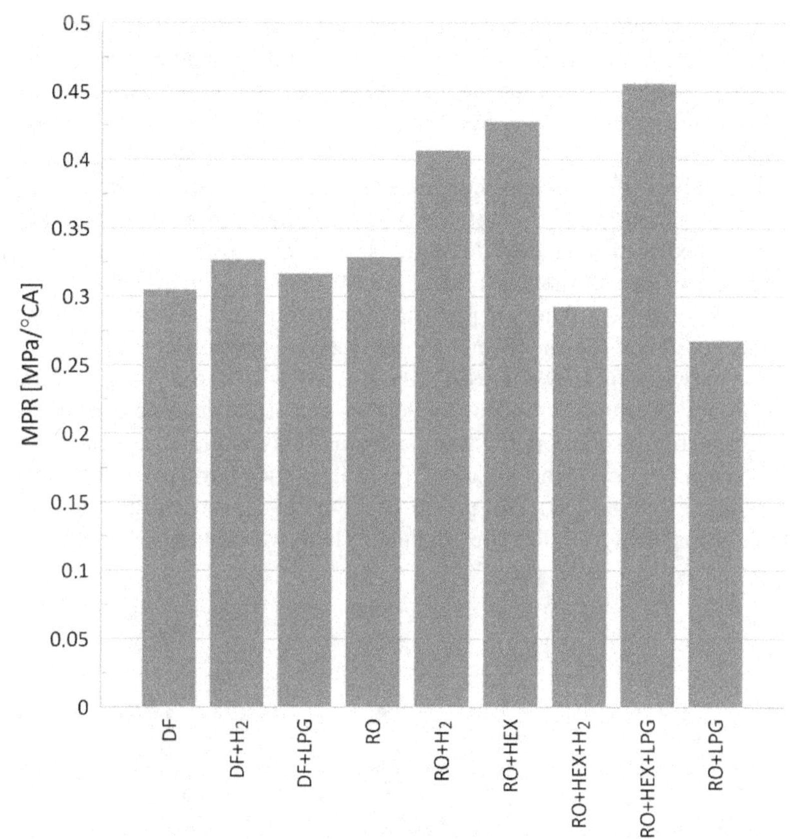

FIGURE 6.3 Maximum combustion pressure rise MPR, with static conditions, vehicle speed 50 km/h, and gear IV

rapeseed oil and hydrogen was very promising. The obtained values of mean indicated pressure were slightly higher than for diesel fuel. The lowest values of mean indicated pressure were obtained for the mixtures of rapeseed oil and hexane burned together with hydrogen. Large values of the mean indicated pressure occurred when rapeseed oil was burned simultaneously with hexane and propane-butane gas and when rapeseed oil was burned simultaneously with propane-butane gas (0.62 MPa). The trends in changes in mean indicated pressure were similar to those observed for engine maximum power and torque. The values of mean indicated pressure are related by a functional relationship to engine power and torque. Engine power and torque were measured using the dynamic method, and the presented mean indicated pressure was determined under static conditions—hence the slight differences. In summary, it should be said that the differences in the values of the mean indicated pressure were small for individual fuels and amounted to a maximum of about 0.04 MPa. The presented results of calculating the mean indicated pressure were averaged from 300 consecutive cycles of engine operation for a given fuel.

The values of the mean indicated pressure are mainly affected by the size of the positive field under the open indicated graph recorded during engine indexing. The size of this field is influenced by the maximum combustion pressure and the angle of combustion onset. The trend of changes in the maximum combustion pressure for individual fuels was similar to that observed for the mean index pressure. However, the maximum difference between fuels was already about 2 MPa—when fueling with rapeseed oil and diesel with hydrogen. A very important parameter impacting the combustion process is the auto-ignition delay angle of the fuel shown in Figure 6.4. It was the smallest for diesel fuel burned with hydrogen and propane-butane gas. The use of other fuels caused an increase in the auto-ignition delay angle. The highest values were obtained for the case

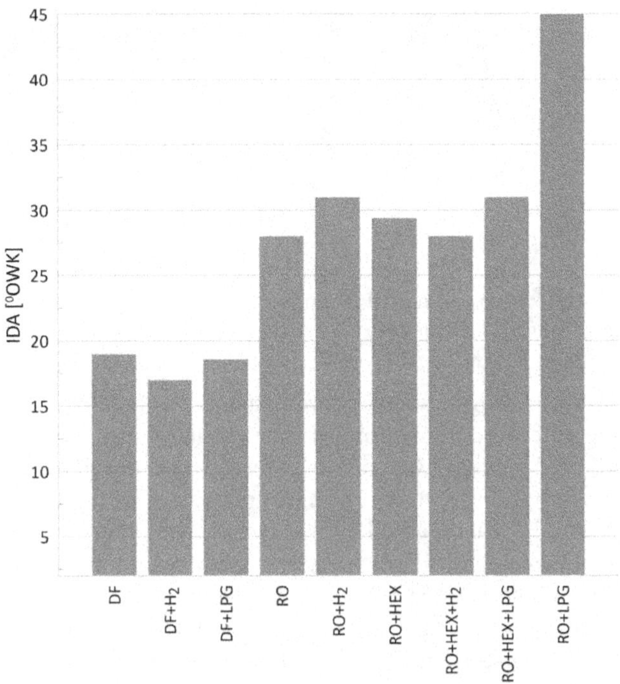

FIGURE 6.4 Auto-ignition delay angle IDA, with static conditions, vehicle speed 50 km/h, and gear IV

of burning rapeseed oil with propane-butane gas. In this case, the auto-ignition delay angle was more than double of that for diesel fuel. Such large differences in the values of the auto-ignition delay angle did not cause significant increases in the maximum pressure rise. The highest values of the maximum pressure build-up velocity were obtained for the combustion of rapeseed oil with hexane and propane-butane gas, as well as for the cases of feeding rapeseed oil with hydrogen and rapeseed oil with hexane—see Figure 6.3. For the other fuels, the value of this parameter was at the level of 0.32 MPa/°CA. The values of the maximum pressure rise are influenced not only by the auto-ignition delay angle but also by the organization of the injection process. This parameter is indicative of the mechanical loads on the structural nodes of the engine. It is desirable to keep its value at a level similar to that of diesel fuel.

6.3 THE COURSE OF HEAT RELEASE AND THE AMOUNT OF HEAT RELEASED

Analyzing in more depth the course of the combustion process in an engine fueled by the tested alternative fuels, it is necessary to determine the course of heat generation and the amount of heat released. These calculations were carried out using the AVL IndiCom program. The results of the calculations in this regard are shown in Figures 6.5 through 6.13. These figures show the heat development course and the amount of heat released calculated from the engine's indicating results. The cycles of engine operation were selected for which the mean index pressures and speed were the most close to the average values of 300 consecutive engine cycles. A number of factors influence the course of heat dissipation. The main ones are the physicochemical properties of the fuel and the injection strategy. The injection strategy will be discussed in Section 6.4. From the point of view of combustion theory, the aim should be that:

- combustion proceeded near the top dead center of the piston to achieve high thermal efficiency;
- the propensity for rapid heat release at the beginning of combustion, induced by the auto-ignition delay, has been maintained at a level that prevents the occurrence of high values of pressure rise; and
- combustion occurred in the entire volume of the combustion chamber.

In the course of the heat release process, we can distinguish three characteristic periods—the period of kinetic (explosive) combustion, the period of

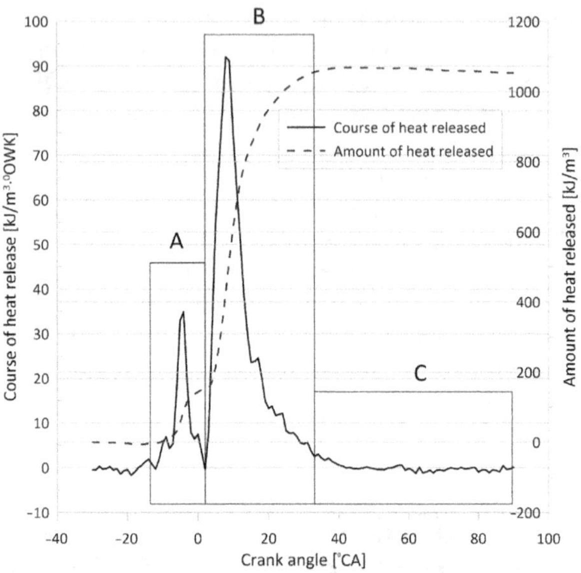

FIGURE 6.5 Course of heat release and amount of heat released, with vehicle speed 50 km/h, gear IV, and diesel fuel; where A—period of kinetic combustion, B—period of diffusion combustion, and C—period of afterburning

diffusion combustion, and the period of afterburning. These periods are shown indicatively in Figure 6.4. An analysis of the presented courses allows us to conclude that:

- for diesel fuel and cases of simultaneous combustion of diesel fuel with hydrogen or propane-butane gas, periods of kinetic combustion, diffusion combustion, and afterburning were clearly observed;
- in cases of fueling with rapeseed oil and its combinations with hexane and gaseous fuels, no clear period of kinetic combustion was observed—the onset of combustion occurred after the top dead center of the piston;
- in the cases of fueling with rapeseed oil and its combinations with hexane and gaseous fuels, the maximum values on the heat release course occurred later after the top dead center of the piston compared to the case of fueling with diesel;
- in the case of the rapeseed oil feed, the maximum values on the heat release curve were more than twice as high as in the case of the diesel feed, but the maximum was located 10°CA farther past the top dead center of the piston than for diesel.

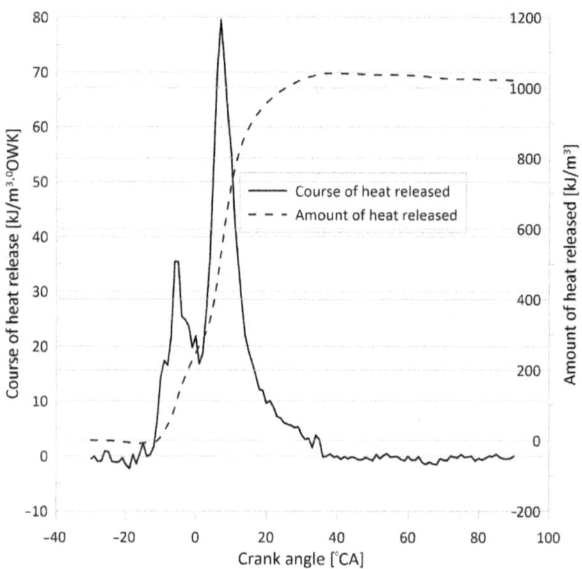

FIGURE 6.6 Heat release course and amount of heat released, with vehicle speed 50 km/h, gear IV, and diesel fuel with hydrogen (DF+H_2)

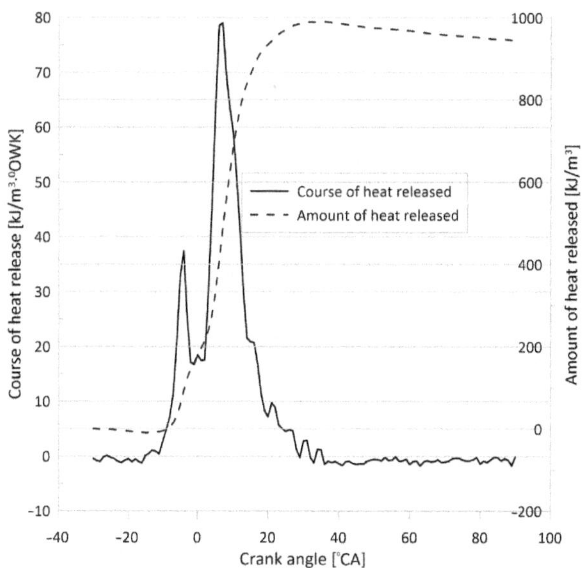

FIGURE 6.7 Course of heat release and amount of heat released, with vehicle speed 50 km/h, gear IV, and diesel fuel and propane-butane gas (DF+LPG)

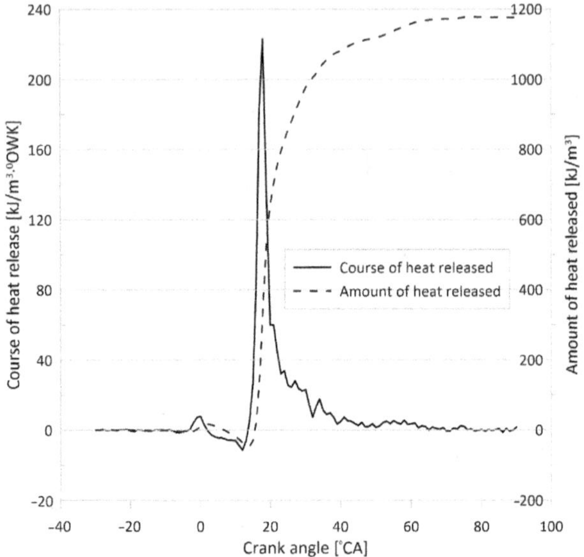

FIGURE 6.8 Heat development course and amount of heat released, with vehicle speed 50 km/h, gear IV, and rapeseed oil (RO)

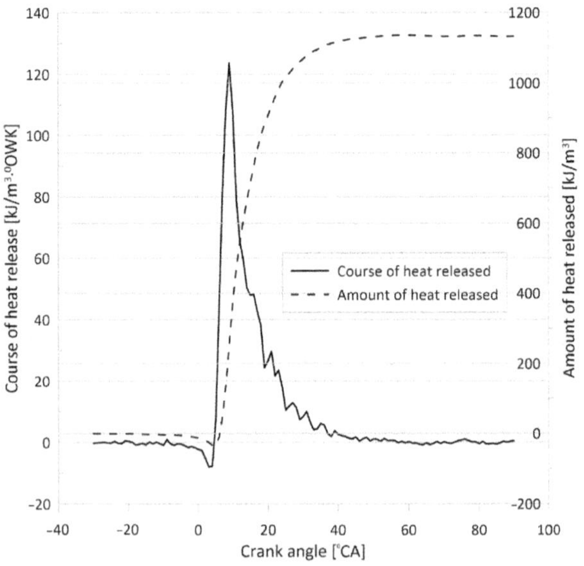

FIGURE 6.9 Course of heat release and amount of heat released, with vehicle speed 50 km/h, gear IV, and rapeseed oil with hydrogen ($RO+H_2$)

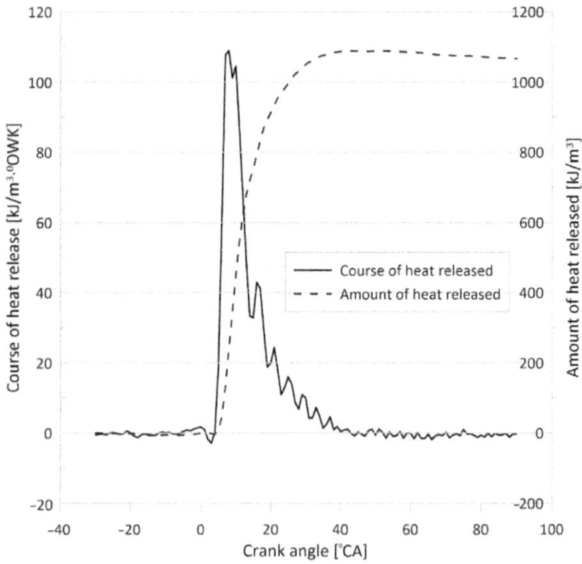

FIGURE 6.10 Course of heat release and amount of heat released, with vehicle speed 50 km/h, gear IV, and rapeseed oil with hexane (RO+HEX)

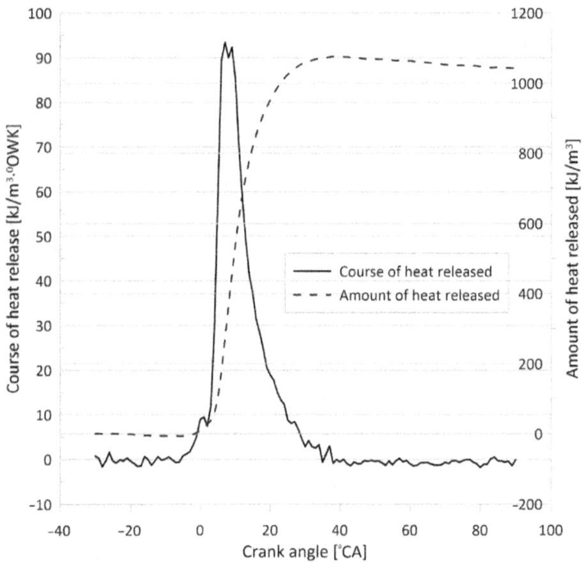

FIGURE 6.11 Course of heat release and amount of heat released, with vehicle speed 50 km/h, gear IV, and rapeseed oil with hexane and hydrogen (RO+HEX+H_2)

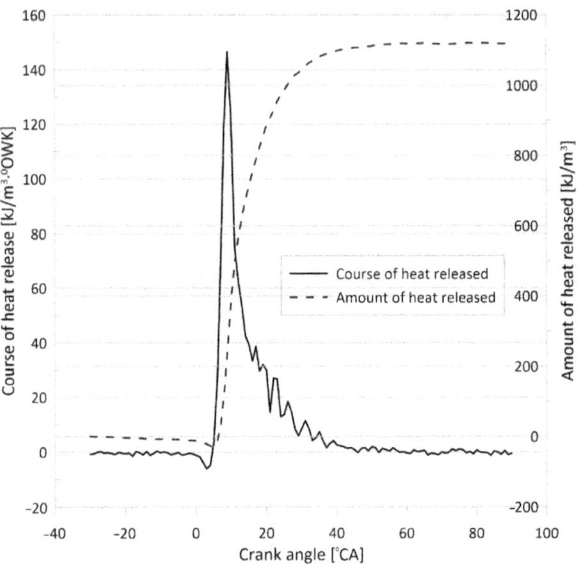

FIGURE 6.12 Course of heat release and amount of heat released, with vehicle speed 50 km/h, gear IV, and rapeseed oil with hexane and propane-butane gas (RO+HEX+LPG)

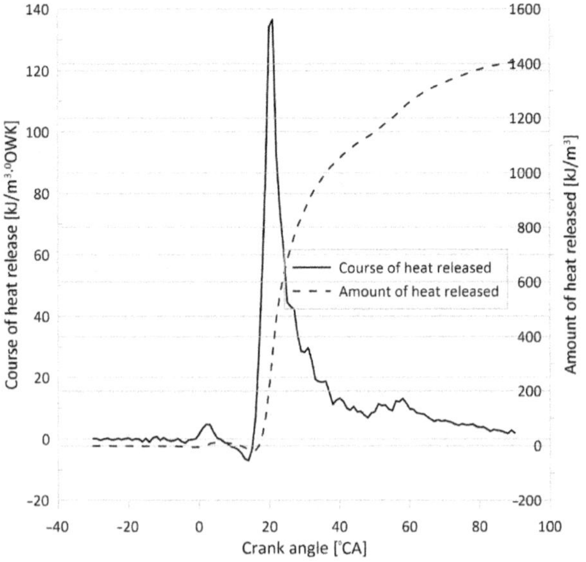

FIGURE 6.13 Course of heat release and amount of heat released, with vehicle speed 50 km/h, gear IV, and rapeseed oil with propane-butane gas (RO+LPG)

6.4 THE COURSE OF THE INJECTION PROCESS

The discussed parameters of the combustion process are mainly influenced by the organization of the fuel injection process. Figure 6.14 shows the volumetric dose of fuel injected per engine cycle as read from the EOBD system. Shown are cases of engine operation under static conditions corresponding to vehicle speeds of 50 km/h and 90 km/h and under dynamic conditions. Static conditions were realized according to points 2 and 3 of the test methodology (Section 4.4). Dynamic conditions were

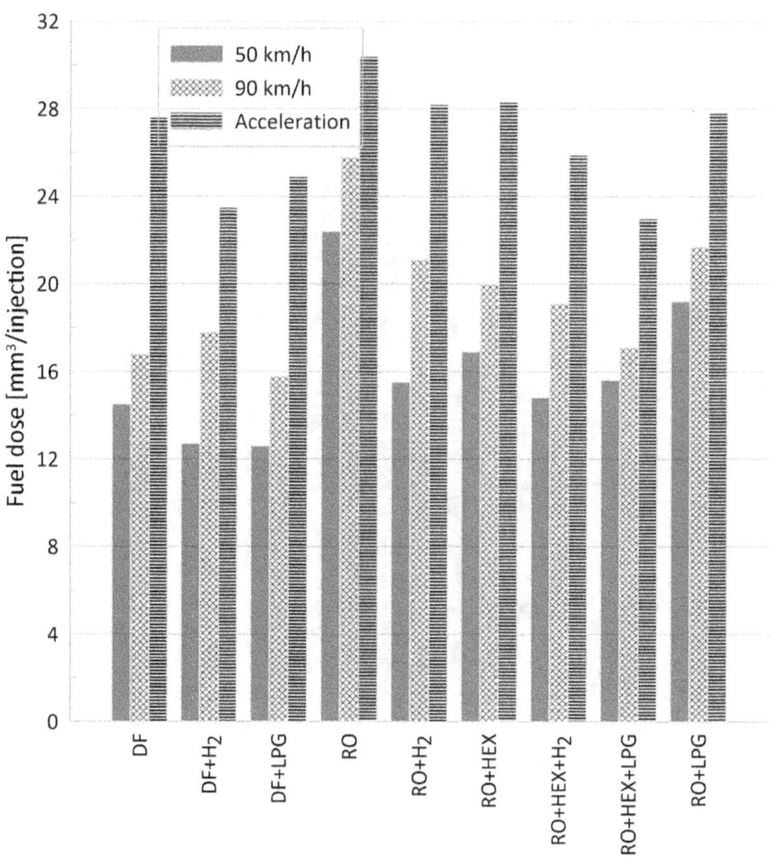

FIGURE 6.14 Liquid fuel dose for single injection, with vehicle speed 50 km/h and gear IV

realized according to point 1 of the test methodology (Section 4.4). The fuel dose under dynamic conditions was averaged among all engine cycles running in the acceleration process. The fuel dose decreased in cases of simultaneous combustion of liquid and gaseous fuels. The diesel quantity control system reduced the liquid fuel dose because additional energy was supplied in the form of gaseous fuels.

The fuel dose recorded under acceleration conditions was much higher than that for static conditions. This is because under dynamic conditions, the vehicle's engine was additionally loaded by inertia forces from the rotating masses of the engine and the vehicle's drivetrain.

Figure 6.15 shows the fuel pressure in the accumulator of the common rail system. This pressure affects the dose of injected fuel and depends on

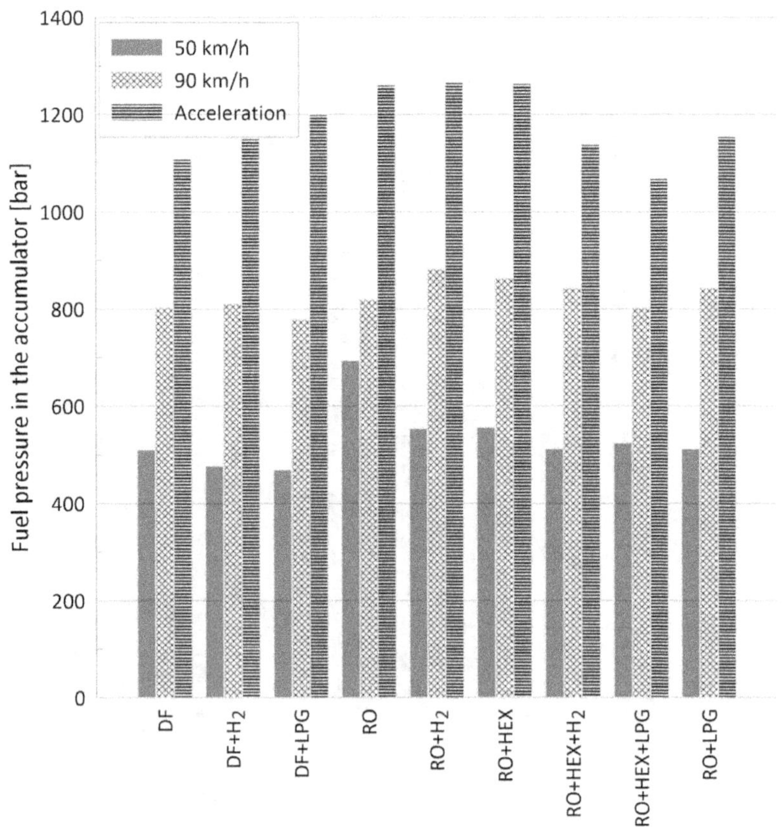

FIGURE 6.15 Liquid fuel pressure in the common rail system tray, with vehicle speed 50 km/h and gear IV

the physicochemical properties of the fuel—mainly viscosity. There is a clear trend in the increase in tray pressure when using rapeseed oil. Rapeseed oil has a significantly higher viscosity than diesel fuel. The addition of hexane to rapeseed oil caused a decrease in the values of pressures in the tray.

Figures 6.16 to 6.24 show the courses of the digital injection signal for the fuels studied, at 50 km/h and gear IV—according to point 2 of the test methodology (Section 4.4). For all of the fuels considered, the engine controller implemented two injections—initial and main. In the case of simultaneous combustion of diesel and gaseous fuels, the initial injection occurred slightly earlier than for diesel. The injection duration angles were very similar. For rapeseed oil, the initial injection occurred about 2°OWK later than for diesel. The end of injection extended significantly until about 12°OWK after the top dead center of the piston. The injection angle was significantly greater than for diesel. The addition of hexane and the simultaneous combustion of rapeseed oil with gaseous fuels resulted in injection parameters similar to those

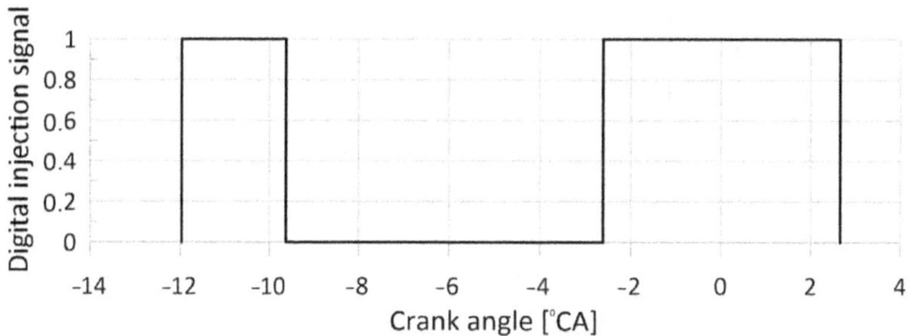

FIGURE 6.16 Digital signal course of liquid fuel injection, with vehicle speed 50 km/h, gear IV, and diesel fuel (DF)

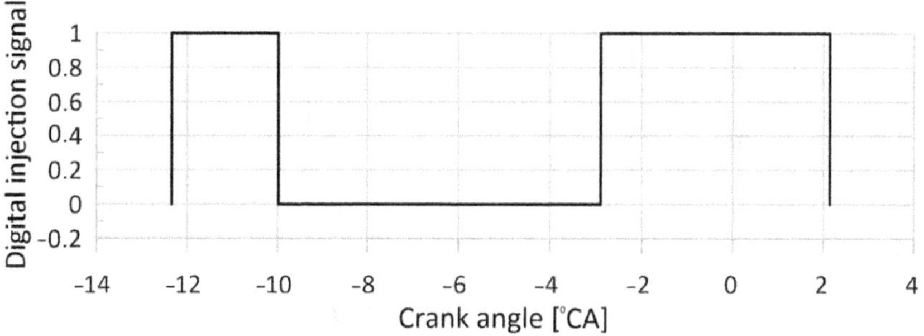

FIGURE 6.17 Signal course of digital liquid fuel injection, with vehicle speed 50 km/h, gear IV, and diesel with hydrogen (DF+H_2)

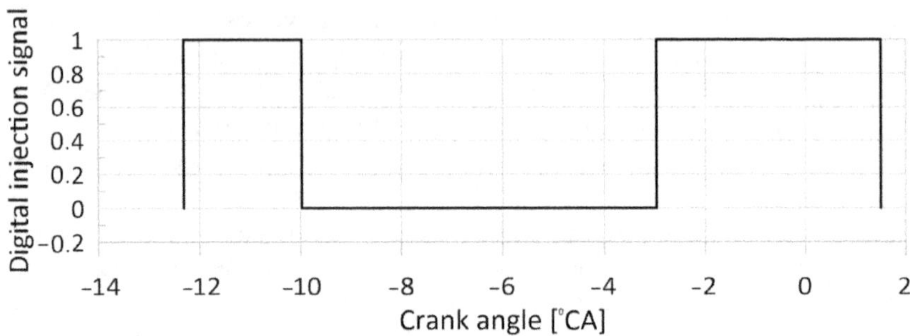

FIGURE 6.18 Digital signal course of liquid fuel injection, with vehicle speed 50 km/h, gear IV, and diesel with propane-butane gas (DF+LPG)

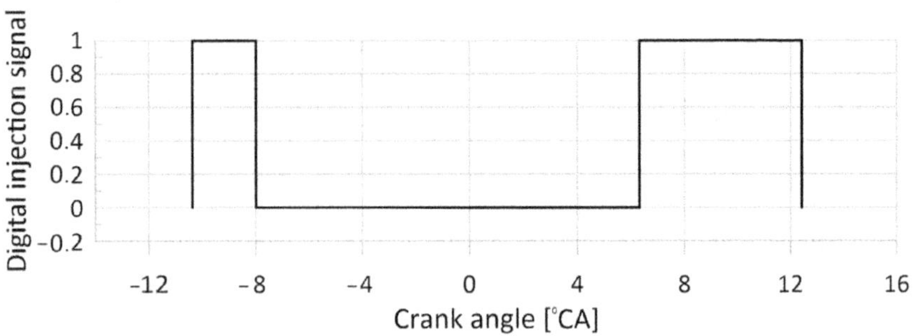

FIGURE 6.19 Digital signal course of liquid fuel injection, with vehicle speed 50 km/h, gear IV, and rapeseed oil RO

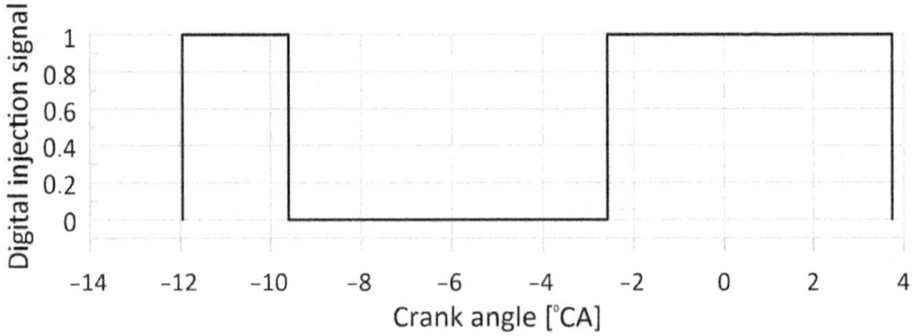

FIGURE 6.20 Digital signal course of liquid fuel injection, with vehicle speed 50 km/h, gear IV, and rapeseed oil with hydrogen ($RO+H_2$)

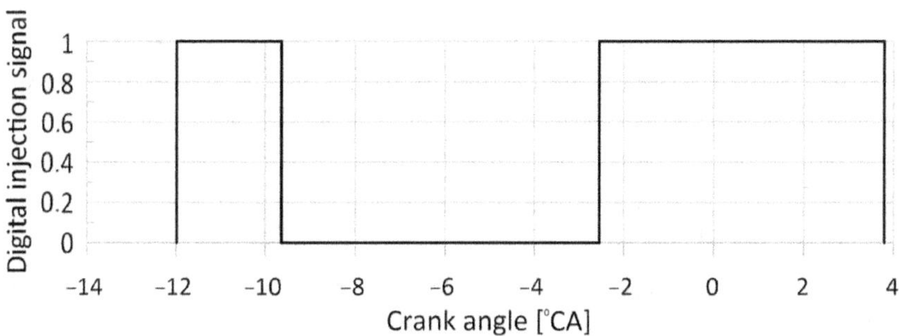

FIGURE 6.21 Signal course of digital liquid fuel injection, with vehicle speed 50 km/h, gear IV, and rapeseed oil with hexane (RO+HEX)

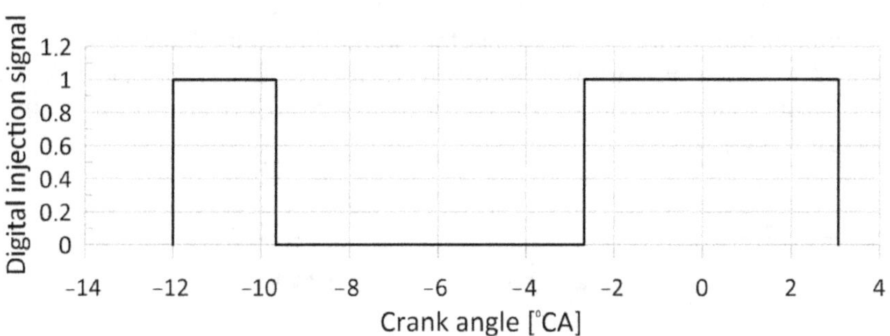

FIGURE 6.22 Signal course of digital liquid fuel injection, with vehicle speed 50 km/h, gear IV, and rapeseed oil with hexane and hydrogen (RO+HEX+H_2)

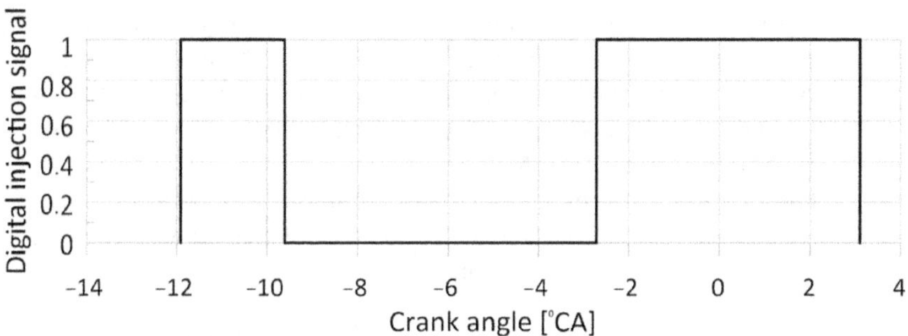

FIGURE 6.23 Digital signal course of liquid fuel injection, with vehicle speed 50 km/h, gear IV, and rapeseed oil with hexane and propane-butane gas (RO+HEX+LPG)

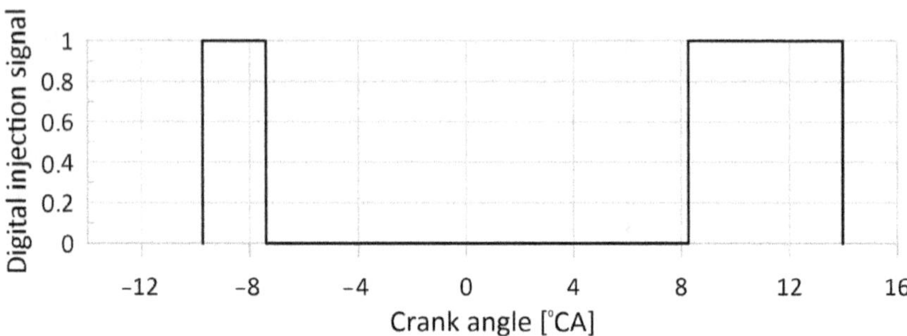

FIGURE 6.24 Digital signal course of liquid fuel injection, with vehicle speed 50 km/h, gear IV, and rapeseed oil with propane-butane gas (RO+ LPG)

for diesel. When fed with rapeseed oil and propane-butane gas, the liquid fuel injection process was similar to that for rapeseed oil. Simultaneous combustion of rapeseed oil with hydrogen without the addition of hexane resulted in a similar injection course as for diesel. Only the end of the main injection occurred slightly later.

6.5 CONCLUSION

On the basis of the study of the course of combustion and injection in a diesel engine, when fueled with alternative fuels, the following were found:

- The trends of changes in average index pressure were similar for those observed for maximum engine power and torque—the values of average index pressure are related by a functional relationship to engine power and torque; the differences in the values of average index pressure were small for individual fuels, amounting to a maximum of about 0.04 MPa.
- The trend of changes in maximum combustion pressure for individual fuels was similar to that observed for average index pressure; however, the maximum difference between fuels was already about 2MPa.
- An important parameter impacting the combustion process is the auto-ignition delay angle of the fuel; it was smallest for diesel fuel burned with hydrogen and propane-butane gas; the use of other fuels increased the auto-ignition delay angle.

- The highest values of the maximum rate of pressure buildup were obtained for the combustion of rapeseed oil with hexane and propane-butane gas, as well as for the cases of feeding rapeseed oil with hydrogen and rapeseed oil with hexane; for the other fuels, the value of this parameter was at 0.32 MPa/°CA.
- For diesel fuel and cases of simultaneous combustion of diesel fuel with hydrogen or propane-butane gas, periods of kinetic combustion, diffusion combustion, and afterburning were clearly observed.
- In cases of fueling with rapeseed oil and its combinations with hexane and gaseous fuels, no clear period of kinetic combustion was observed—the onset of combustion occurred after the top dead center of the piston.
- For all of the fuels considered, the engine controller implemented two injections—the initial and main injection; in the case of simultaneous combustion of diesel with gaseous fuels, the initial injection occurred slightly earlier than for diesel; the injection angles were very similar; for rapeseed oil, the initial injection occurred about 2°OWK later than for diesel; the end of the injection extended significantly until about 12°OWK after the top dead center of the piston; the injection angle was significantly greater than for diesel; the addition of hexane and the simultaneous combustion of rapeseed oil with gaseous fuels resulted in injection parameters similar to those for diesel; when fed with rapeseed oil and propane-butane gas, the liquid fuel injection course was similar to that for rapeseed oil; the simultaneous combustion of rapeseed oil with hydrogen without the addition of hexane resulted in a similar injection course to that for diesel—only the end of the main injection occurred slightly later.

Emissions of Toxic Exhaust Constituents

7

7.1 INTRODUCTION

Pollutants emitted from diesel engines contribute to the formation, greenhouse effect, smog, and acid rain. This affects the health of the population—mainly in cities. The products of combustion of diesel fuel (hydrocarbon fuels) in diesel engines are carbon dioxide, hydrogen, nitrogen, and oxygen. Among other things, the incomplete combustion of hydrocarbon fuels additionally produces carbon monoxide, hydrocarbons, nitrogen oxides, and particulates. More than 90% of the aerosol particles produced by diesel combustion have an aerodynamic diameter of less than 1 μm. The amount of particles emitted is influenced by a number of factors, including the physicochemical properties of the fuel, the combustion process, injection, and the preparation of the combustible mixture. Diesel engine exhaust contains 20 to 100 times more PM than petrol engine exhaust. Due to the differing physical and chemical properties of particulate matter, its exact characterization is difficult. The structure and chemical composition of PM depend on operating conditions, engine type and condition, fuel type, fuel additives and engine lubricating oil, operating conditions, and emission control systems. Particulate matter varies in shape and size, typically consisting of individual soot particles and larger agglomerates. Elemental carbon is the main component of particulate matter on whose surface organic and inorganic compounds, mainly sulfates, adsorb. Particulates with very small diameters are particularly dangerous and can have mutagenic and carcinogenic effects when inhaled by humans. Diesel exhaust is dominated by particulates with a diameter of 0.02–0.5 μm,

DOI: 10.1201/9781003466291-7

which can agglomerate to a diameter of up to 30 μm. Carbon monoxide is a colorless, odorless, and tasteless gas, mainly formed by the insufficient combustion of carbon-containing substances. Carbon dioxide (CO_2), on the other hand, is formed by efficient combustion and is not a toxic gas. The emission of large amounts of CO_2 on a global scale disrupts the thermodynamic balance of the atmosphere, leading to the so-called greenhouse effect. Hydrocarbons (HC) are toxic chemical compounds with negative effects on human health. They are compounds consisting of carbon (C) and hydrogen (H), which are found in large quantities in crude oil and natural gas. The main cause of hydrocarbon emissions from engines is the cooling effect of the combustion chamber walls. The extract of the organic part of the particulate matter mainly contains unburned hydrocarbons that were originally in the fuel. The organic part of the soluble fraction consists of an aromatic phase, derived from unburned hydrocarbons, and a paraffinic, acidic (e.g., cresol, benzoic acid), and basic (e.g., pyridine, aniline) phase. Hydrocarbons are also formed in exhaust gas depletion reactions as a result of slow combustion. Nitrogen oxides emitted by compression-ignition engines are referred to as NO_x. Photochemical reactions involving nitrogen oxides can lead to smog formation. Under conditions of high temperature and pressure, nitrogen reacts with oxygen to form nitric oxide (NO) and, in lesser amounts, nitrogen dioxide (NO_2) and nitrous oxide (N_2O). The majority of nitrogen oxides in diesel exhaust come from the oxidation of nitrogen contained in the air. The intensive formation of NO occurs early in the combustion process due to the high temperature, pressure, and long reaction time. Current technologies use various exhaust gas aftertreatment systems, reducing the harmful effects on the atmosphere. In the present study, flue gases were taken for analysis from the mentioned flue gas cleaning systems. The presented research results should be regarded as preliminary. This is because of the measuring system used.

7.2 TEST RESULTS

Figures 7.1 to 7.5 show the concentrations of exhaust components emitted from the diesel engine when fueled with the fuels tested. Figure 7.1 shows the concentration of carbon monoxide. Figure 7.2 shows the concentration of carbon dioxide. The maximum absolute differences in carbon monoxide concentration were about 0.23% for a vehicle speed of 50 km/h with the fuels DF+H_2 and RO+LPG. The highest carbon monoxide concentrations were

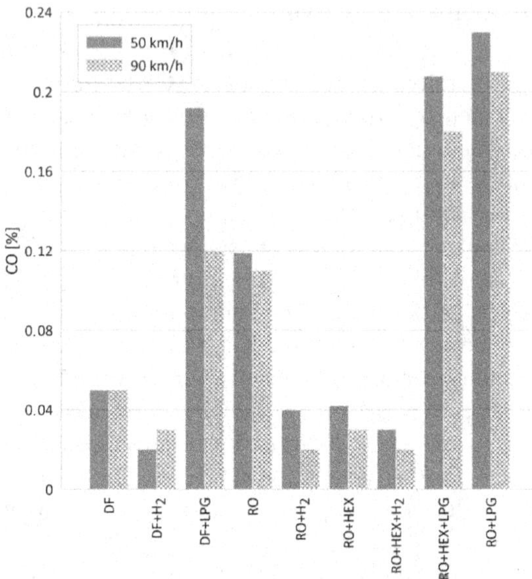

FIGURE 7.1 Carbon monoxide emission concentration in the exhaust before exhaust aftertreatment system is used, with vehicle speed 50 km/h and gear IV

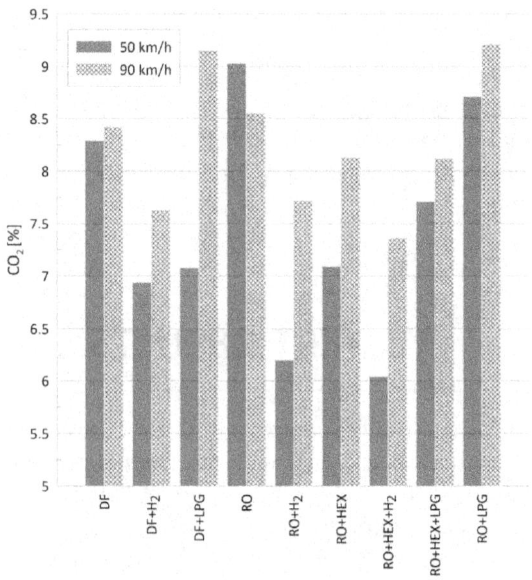

FIGURE 7.2 Concentration of carbon dioxide emissions in exhaust gas before the aftertreatment system is used, with vehicle speed 50 km/h and gear IV

obtained for DF+LPG, RO, RO+HEX+LPG, and RO+LPG fuels. The use of hydrogen as an additional fuel in the combustion of diesel and rapeseed oil resulted in lower carbon monoxide concentrations than for diesel. A similar phenomenon occurred for carbon dioxide concentrations. Figure 7.3 shows the hydrocarbon concentrations. The hydrocarbon concentration for diesel combustion was around 30 ppm. The hydrocarbon concentration values for fueling diesel with hydrogen, rapeseed oil with hydrogen, rapeseed oil with hydrogen and rapeseed oil with hexane were at similar levels. The highest values of hydrocarbon concentrations were obtained for DF+LPG, RO, RO+HEX+LPG, and RO+LPG fuels. Figure 7.4 shows the concentration of nitrogen oxides in the exhaust gas. The lowest values of nitrogen oxide concentrations were obtained for fuels with high carbon monoxide concentrations—which is consistent with the theory of nitrogen oxide formation. The averaged differences between diesel and other fuels were around 200 ppm. Figure 7.5 shows the smoke opacity levels for the fuels tested. The highest smoke opacity values were found for rapeseed oil and rapeseed oil combusted with propane-butane gas. Figure 7.6 shows the excess air factor l and the oxygen concentration in the exhaust gas before the system exhaust gas aftertreatment system.

The ratio λ defines the ratio between the amount (volume) of air actually supplied for combustion and the amount of air theoretically needed to

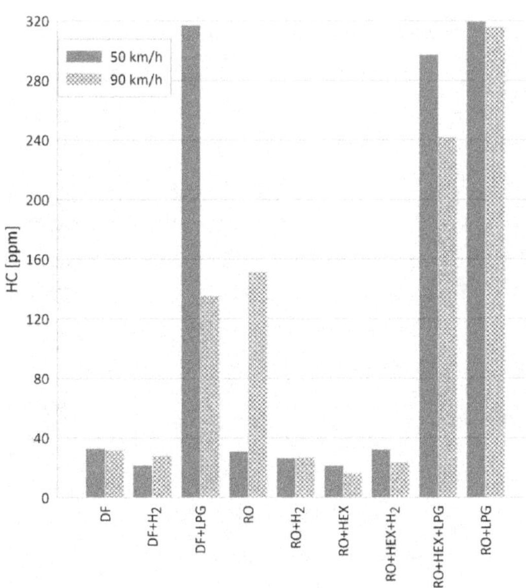

FIGURE 7.3 Concentration of hydrocarbon emissions in the exhaust gas before the aftertreatment system is used, with vehicle speed 50 km/h and gear IV

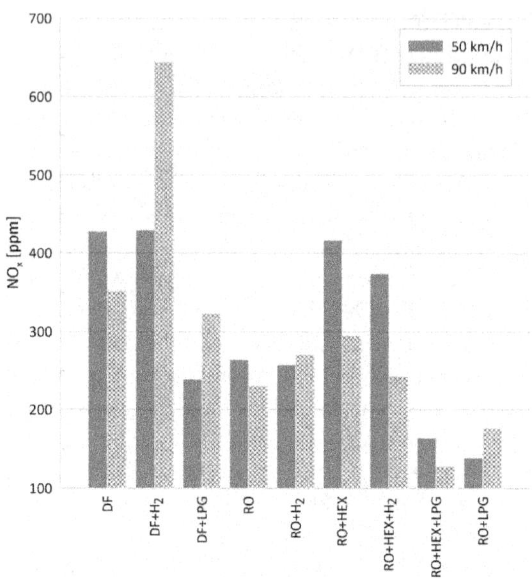

FIGURE 7.4 Concentration of nitrogen oxide emissions in the exhaust gas before the aftertreatment system is used, with vehicle speed 50 km/h and gear IV

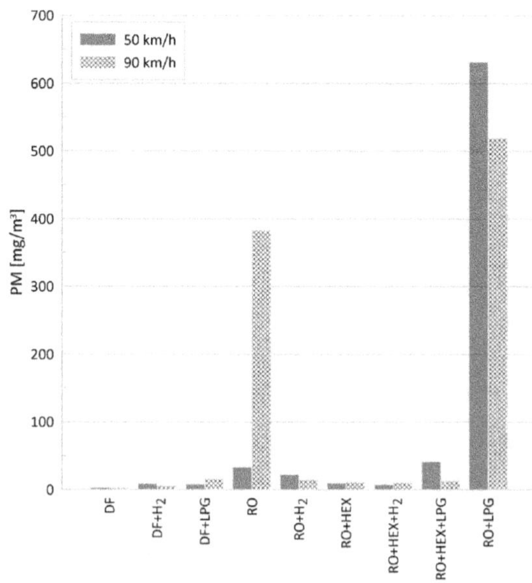

FIGURE 7.5 Smoke opacity before the exhaust aftertreatment system is used, with vehicle speed 50 km/h and gear IV

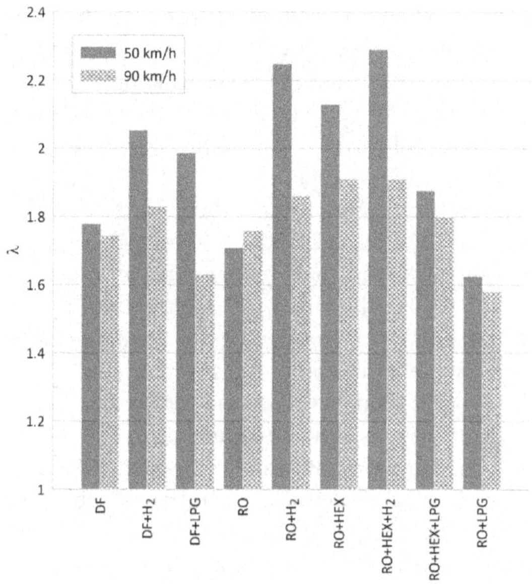

FIGURE 7.6 Exhaust gas excess flow rate before the exhaust aftertreatment system is used, with vehicle speed 50 km/h and gear IV

completely and totally burn the fuel contained in the combustion mixture. The excess air ratio has a large effect on the combustion temperature, which translates into the amount of fuel consumed. The higher the excess air ratio, the lower the combustion temperature. For the combustion of rapeseed oil and rapeseed oil with propane-butane gas, an increased combustion temperature was observed as compared to diesel. The excess air ratio was the lowest for these fuels.

7.3 CONCLUSION

On the basis of tests carried out on the concentration of toxic emissions upstream of the aftertreatment system, when fueled with alternative fuels, it was concluded that

- the highest carbon monoxide concentrations were obtained for DF+LPG, RO, RO+HEX+LPG, and RO+LPG; the use of hydrogen as an additional fuel in the combustion of diesel and rapeseed oil

resulted in lower carbon monoxide concentrations than for diesel—a similar phenomenon occurred for carbon dioxide concentrations;
- hydrocarbon concentrations in the combustion of diesel were around 30 ppm; hydrocarbon concentrations in the simultaneous combustion of diesel with hydrogen, rapeseed oil with hydrogen, rapeseed oil with hydrogen, and rapeseed oil with hexane were at similar levels; the highest hydrocarbon concentrations were obtained for DF+LPG, RO, RO+HEX+LPG, and RO+LPG;
- the lowest values for nitrogen oxide concentrations were obtained for fuels with high carbon monoxide concentrations—which is consistent with the theory of nitrogen oxide formation; averaged differences between diesel and other fuels were around 200 ppm;
- the highest smoke values were found for rapeseed oil and rapeseed oil burnt together with propane-butane gas;
- emission concentrations of toxic components downstream of the aftertreatment system were very similar for the fuels tested; of course, differences in emission concentrations upstream of the aftertreatment system will affect the effectiveness of the aftertreatment process over a wide range of engine operating conditions and the durability of the aftertreatment system components.

Conditions of Dynamic Engine Operation 8

8.1 INTRODUCTION

The engine of a motor vehicle operates mainly under dynamic conditions. Therefore, the dynamic operating conditions of the engine are represented in the WLTP test—they form a significant part of the test. Particularly important here is the engine acceleration process caused by a step change in the position of the fuel dosage control lever. These conditions are mainly used by car engines operating in urban conditions. Project development using new fuel types was therefore planned for testing during the engine acceleration process. The operation of an internal combustion engine under dynamic conditions is characterized by certain peculiarities, as discussed in subsection 3.2, Chapter 3. In addition, the different physicochemical properties of the alternative fuels considered in relation to diesel are of significance here. As mentioned in subsection 4.4, Chapter 4, the tests were carried out in Condition no. 1—acceleration test—acceleration of the vehicle in gear II on a chassis dynamometer, the dynamometer rollers rolled freely, and acceleration was initiated by stepping up the lever controlling the fuel dose. During such an acceleration process, the vehicle engine was loaded with the moment of inertia from the rotating masses of the engine and the rotating masses of the drive train associated with the engine crankshaft. In addition, the engine load consisted of the moment of inertia from the rotating masses of the chassis dynamometer. Engine operating cycles running at similar engine crankshaft speeds were selected in order to compare the operational parameters of the engine when fueled with different fuels. It was assumed that comparisons of operational parameters of the engine operation would be carried out at an engine crankshaft speed of 4,000 rpm. The results obtained made it possible to assess the suitability of the fuels tested under dynamic conditions. In the case of engine fueling with liquid and gaseous fuels, the optimal control parameters of the gaseous fuel injection process were determined under static conditions.

8.2 TEST RESULTS

Figure 8.1 shows the course of the engine crankshaft speed for each cycle of the acceleration process. It was found that the course of the engine crankshaft speed varied for each of the fuels considered. The most favorable course of change in engine crankshaft speed occurred for diesel and simultaneous combustion of diesel with hydrogen or propane-butane LPG. The engine fueled with rapeseed oil was the slowest to accelerate. The above phenomenon was reflected in the values of the average angular acceleration calculated for the acceleration process when the engine was fueled with the fuels under consideration. The values of the average angular acceleration are shown in Figure 8.2. The highest acceleration values were obtained when the engine was simultaneously fueled with diesel and LPG propane-butane gas. These were around 68 s^{-2}. The lowest acceleration values were obtained when it was fueled with rapeseed oil and hydrogen. These were around 36 s^{-2}. In general, the highest angular accelerations were obtained when burning diesel and when burning diesel simultaneously with gaseous fuels. The lowest angular acceleration values were obtained when the engine was fueled with rapeseed oil, rapeseed oil with hexane, rapeseed oil with hydrogen, and rapeseed oil with hexane and hydrogen. Some improvement with rapeseed oil was obtained by simultaneously burning rapeseed oil with LPG propane-butane gas. The results shown indicate that, when simultaneously combusting rapeseed oil with hydrogen under dynamic conditions, it is necessary to modify the gaseous fuel control process—this control must be different from that for static conditions. Further work will be carried out in this area.

Figure 8.3 shows an example of the mean indicated pressure of the recorded cycles of the acceleration process, when seeded with diesel. Figure 8.4 shows the averaged values of the mean indicated pressure among all cycles of the acceleration process for individual fuels. The trend of changes in this parameter is similar to the trend of changes in the maximum angular acceleration of the engine crankshaft averaged over the entire acceleration process, for the individual fuels. The highest values of the averaged mean indicated pressure

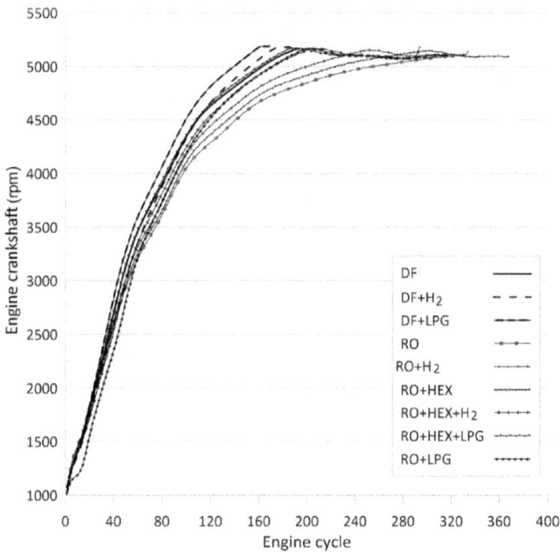

FIGURE 8.1 Course of the engine crankshaft speed for individual cycles of the acceleration process

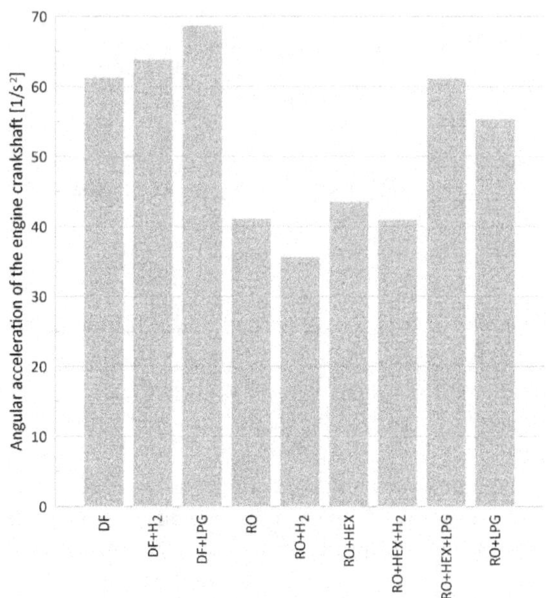

FIGURE 8.2 Maximum angular accelerations of the engine crankshaft averaged over the entire acceleration process

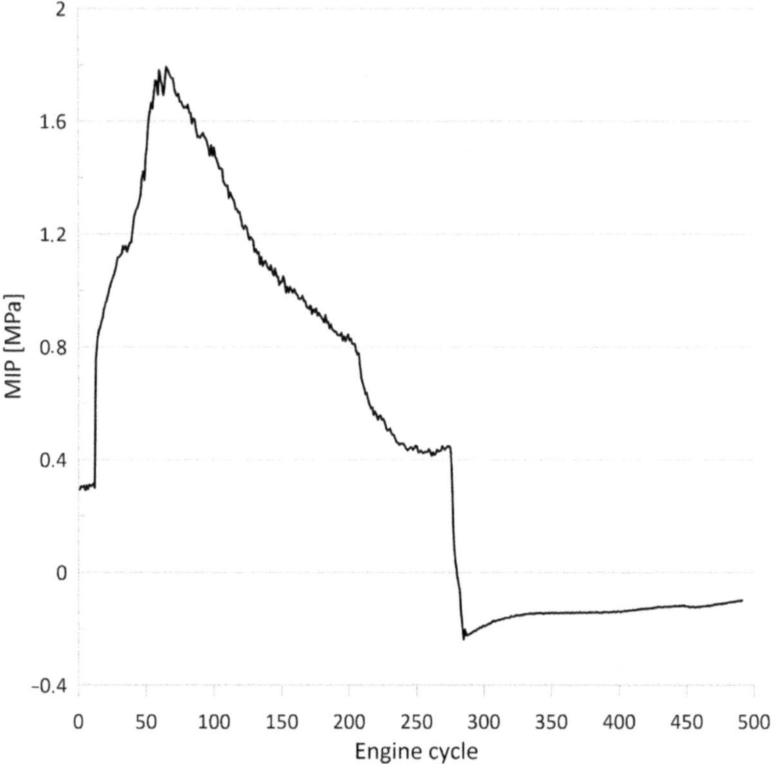

FIGURE 8.3 Example of the average pressure curve for individual cycles of the acceleration process, when the engine is fed with diesel fuel

were obtained when diesel fuel was burned and when diesel fuel was burned simultaneously with gaseous fuels. The lowest values of the averaged mean indicated pressure were obtained when the engine was fueled with rapeseed oil. However, the differences were not significant and amounted to a maximum of about 0.2 MPa.

In the following analyses, selected combustion process parameters (mean indexed pressure—Figure 8.5, maximum combustion pressure—Figure 8.6, and maximum pressure build-up speed—Figure 8.7) were compared for acceleration process cycles running at an engine crankshaft speed of approximately 4,000 rpm. As with the parameters discussed earlier, the mean indicated pressure and maximum combustion pressure were the highest for the case of diesel combustion and simultaneous combustion of diesel with diesel. The maximum

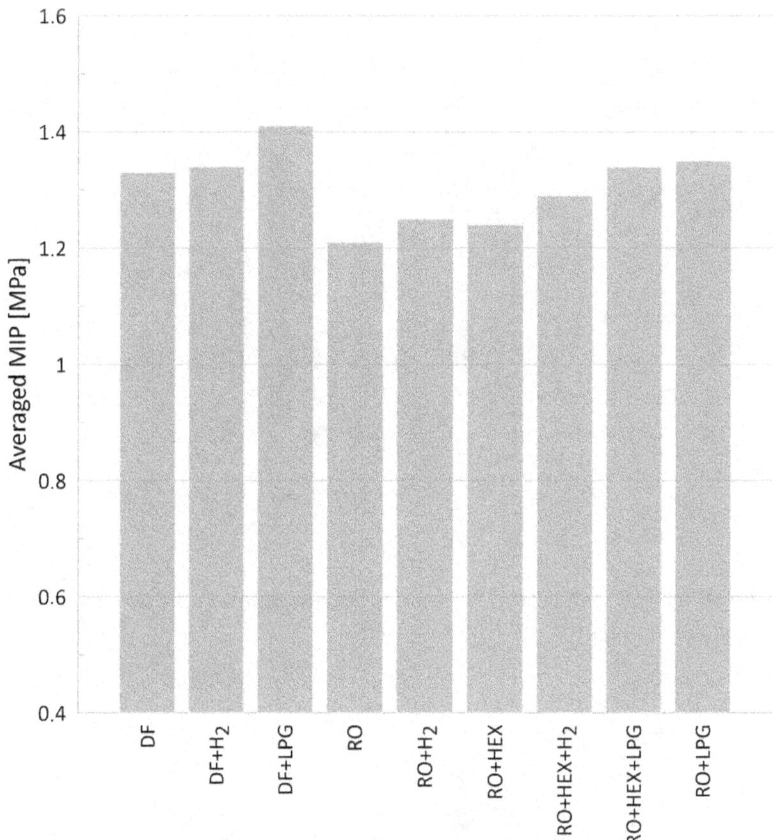

FIGURE 8.4 Mean indicated pressure (MIP) averaged over the entire acceleration process

pressure build-up rate was the smallest for the case of burning rapeseed oil with hydrogen and the highest for the case of burning diesel. The absolute difference was significant at around 2 MPa/°CA. The reason for the observed differences is the chronic combustion process for the rapeseed oil feed cases. This is because the differences in maximum combustion pressures were negligible, with a maximum of around 1 MPa. The auto-ignition delay angle was very similar for the individual feed cases, at around 15°CA. The injection system carried out one fuel injection per engine cycle at an engine crankshaft speed of 4,000 rpm.

Figure 8.8 shows the fuel dose injected per engine cycle when feeding the different fuels. The use of rapeseed oil as a fuel resulted in an increase in

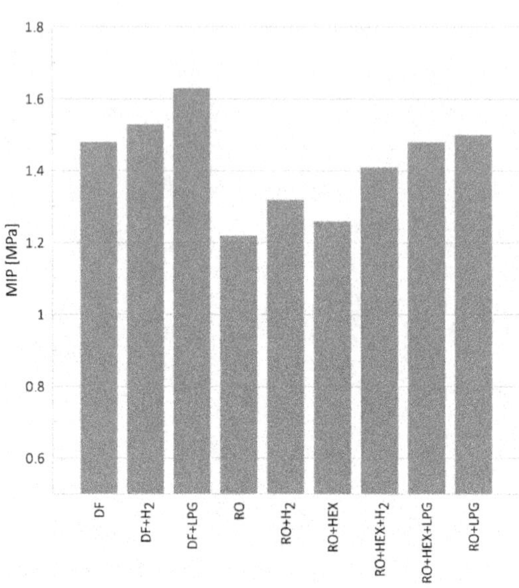

FIGURE 8.5 Mean indicated pressure (MIP), acceleration process, with engine crankshaft speed approximately 4,000 rpm

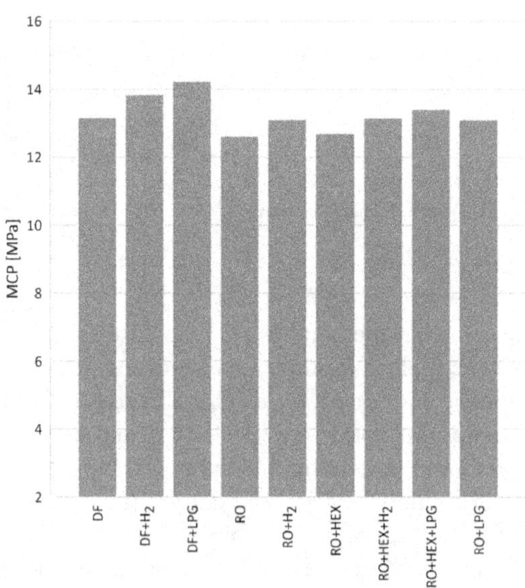

FIGURE 8.6 Maximum combustion pressure (MCP), acceleration process, with engine crankshaft speed approximately 4,000 rpm

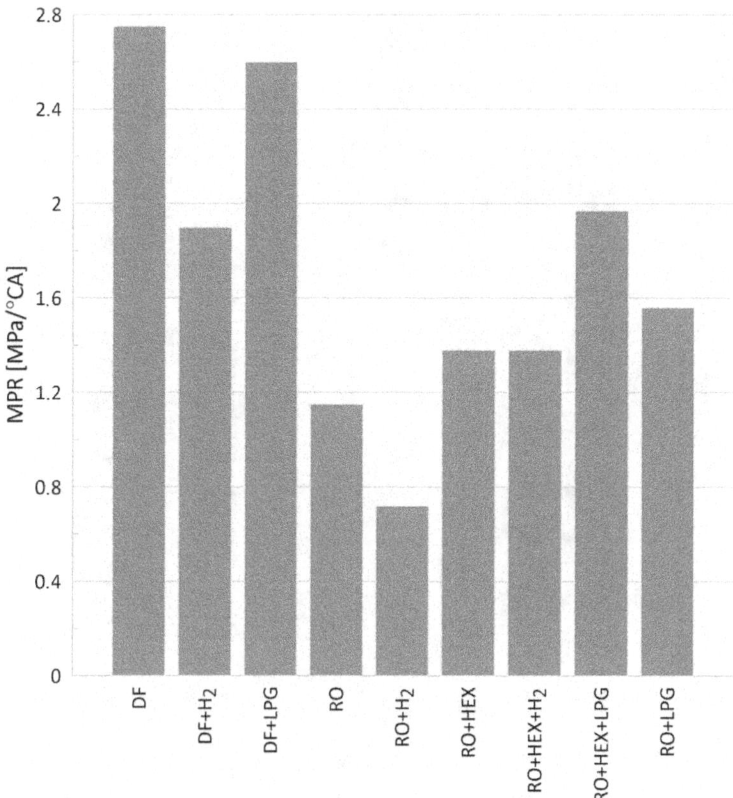

FIGURE 8.7 Maximum combustion pressure rise (MPR), acceleration process, with engine crankshaft speed around 4,000 rpm

the amount of liquid fuel injected per engine cycle. The maximum difference between diesel and rapeseed oil was about 1.5 mm^3 /injection. The injection system, taking into account the parameters recorded by the common rail system sensors, increased the opening time of the injectors.

Figures 8.9 to 8.13 show the heat development and heat release calculated from the engine indexing results. Engine operating cycles were selected for which the engine crankshaft speed was approximately 4,000 rpm. The calculation results are shown for the cases of diesel, diesel with hydrogen, rape seed oil, rape seed oil with hydrogen, and rape seed oil with hexane and hydrogen.

An analysis of the heat release curves and the amount of heat released shows that there was a clear period of diffusion combustion for diesel and the cases of simultaneous combustion of diesel with hydrogen. The situation was

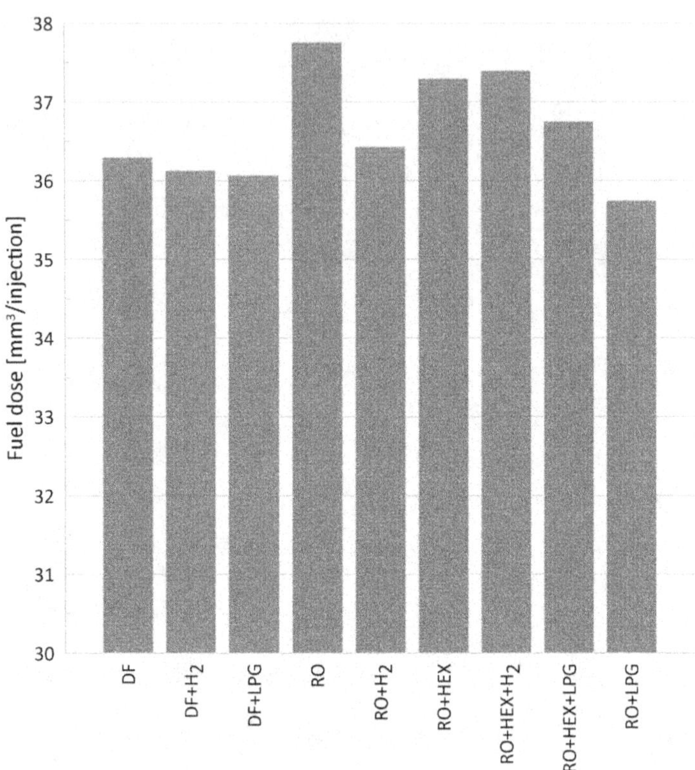

FIGURE 8.8 Liquid fuel dose for single injection, acceleration process, with engine crankshaft speed approximately 4,000 rpm

similar for the combustion of rapeseed oil, rapeseed oil with hydrogen, and rapeseed oil with hexane and hydrogen. The situation was therefore different to that under static conditions, where there was a clear period of kinetic combustion for the diesel fuel cases. The point of onset of heat release was similar for all of the fuels considered. In the case of diesel fuel and simultaneous fueling with diesel and hydrogen, the maximum heat release was about 3,200 kJ/m^3. In the case of rapeseed oil feed, the maximum heat release decreased significantly and was about 2,600 kJ/m^3. This situation improved when the engine was simultaneously fueled with rapeseed oil, hexane, and hydrogen. The maximum heat release increased to about 3,100 kJ/m^3. When the engine was fueled with rapeseed oil and hydrogen, the maximum heat release was located approximately 10°CA further downstream of the top dead center of the piston than for diesel.

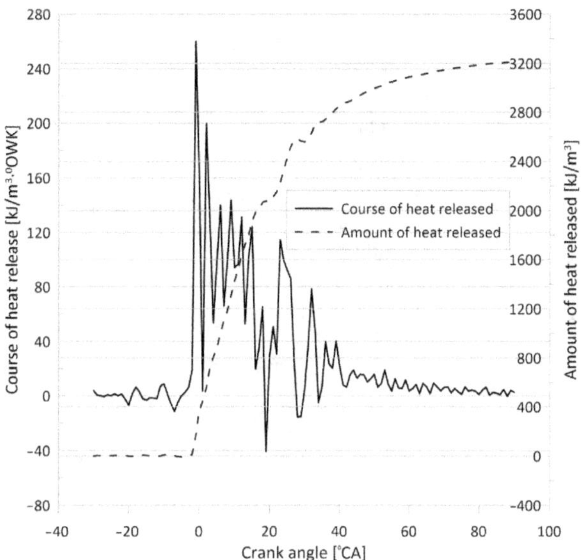

FIGURE 8.9 Heat release course and amount of heat released, acceleration process, with engine crankshaft speed around 4,000 rpm and DF

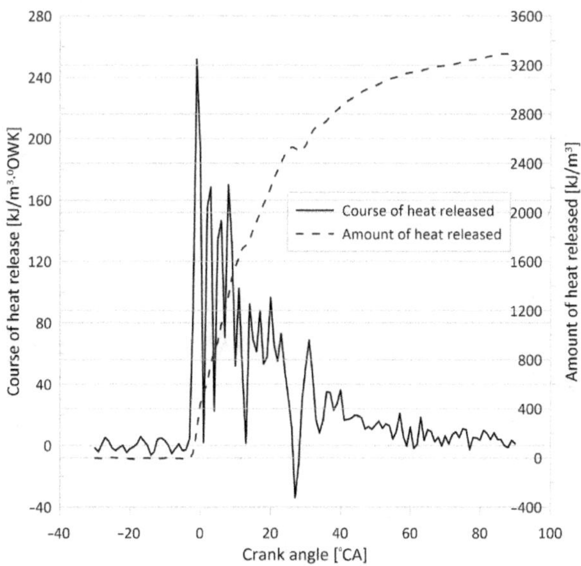

FIGURE 8.10 Heat release course and amount of heat released, acceleration process, with engine crankshaft speed around 4000 rpm and DF+H_2

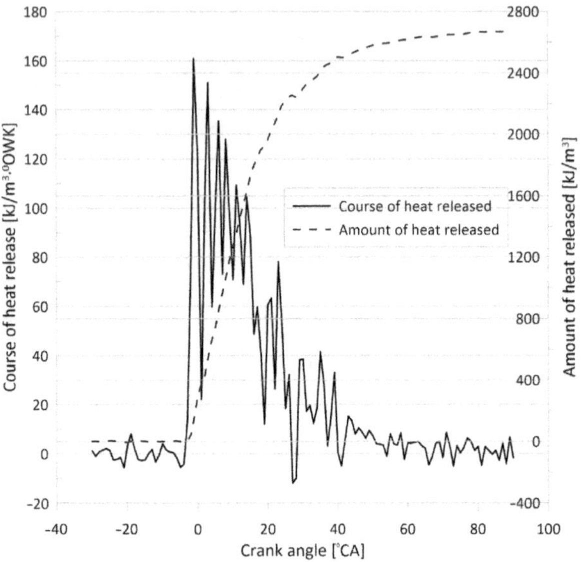

FIGURE 8.11 Heat release course and amount of heat released, acceleration process, with engine crankshaft speed around 4,000 rpm and RO

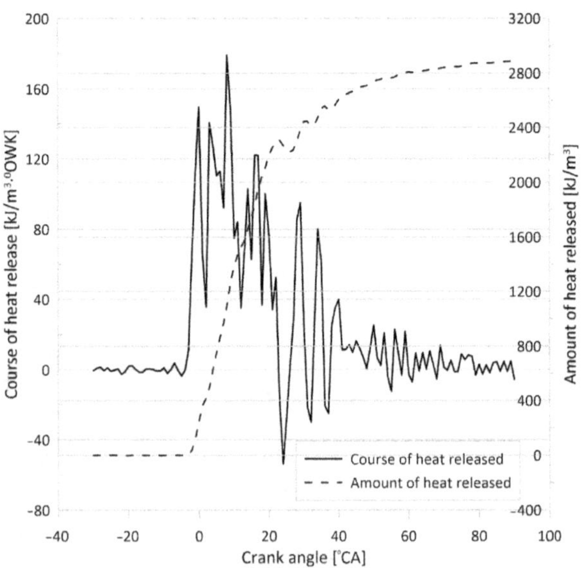

FIGURE 8.12 Heat release course and amount of heat released, acceleration process, with engine crankshaft speed around 4,000 rpm and RO+H_2

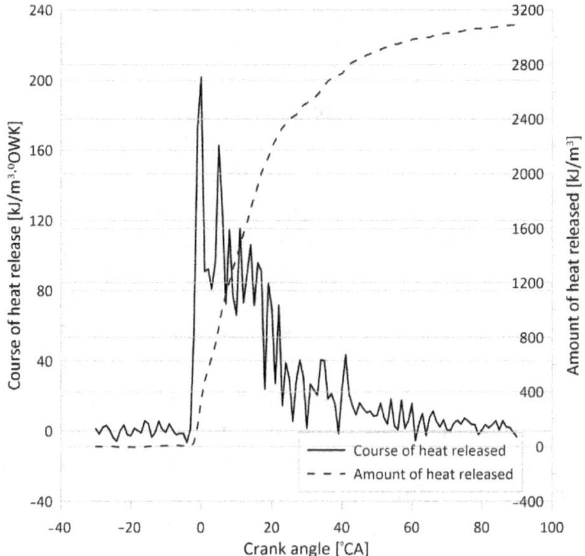

FIGURE 8.13 Heat release course and amount of heat released, acceleration process, with engine crankshaft speed around 4,000 rpm and RO+HEX+H_2

8.3 CONCLUSION

An analysis of the presented waveforms of test results allows us to conclude that:

- the highest values of angular acceleration of the engine crankshaft during the acceleration process were obtained when the engine was simultaneously fueled with diesel and LPG propane-butane gas—they were about 68 s^{-2}; the lowest values of acceleration were obtained when it was fueled with rapeseed oil and hydrogen—they were about 36 s^{-2};
- in the case of simultaneous combustion of rapeseed oil with hydrogen under dynamic conditions, it is necessary to modify the gaseous fuel control process—this control must be different from that for static conditions;
- the highest values of the averaged average index pressure were obtained when burning diesel and when simultaneously burning diesel and gaseous fuels; the lowest values of the averaged average

index pressure were obtained when feeding rapeseed oil—the differences were not significant, however, and amounted to a maximum of around 0.2 MPa;

- the maximum rate of pressure buildup was the lowest for the case of burning rapeseed oil with hydrogen and the highest for the case of burning diesel; the absolute difference was significant at around 2 MPa/°CA. The reason for the observed differences was the chronic combustion process for the rapeseed oil feed cases;
- for diesel and cases of simultaneous combustion of diesel with hydrogen, there was a clear period of diffusion combustion; a similar situation occurred for the combustion of rapeseed oil, rapeseed oil with hydrogen, and rapeseed oil with hexane and hydrogen—the situation was therefore different than under static conditions; and
- in the case of diesel fuel and simultaneous supply of diesel and hydrogen, the maximum heat release was approximately 3,200 kJ/m^3. For the case of feeding with rapeseed oil, the amount of heat released decreased significantly and was approximately 2,600 kJ/m^3.

9 Summary

Today's automotive industry faces major challenges in reducing exhaust emissions and reducing fossil fuel consumption. In the face of growing environmental awareness and tightening regulations, the search for alternative power sources for internal combustion engines has become a priority for both industry and researchers. In particular, diesel engines, which are widely used in commercial vehicles and agricultural and construction machinery, require modern solutions to reduce their negative impact on the environment while maintaining their energy efficiency. Diesel engines, with their high efficiency and long service life, play a key role in road transportation and heavy industry. However, their traditional way of powering them, based mainly on diesel, is associated with emissions of harmful substances such as nitrogen oxides (NOx), particulate matter (PM), and carbon dioxide (CO_2). In response to these challenges, there has been a great deal of research in recent years into the use of alternative fuels that could reduce emissions while helping to diversify energy sources for powering transportation vehicles. Research on alternative fuels for diesel engines has paid a particular attention to several types of fuels that differ both chemically and physically, resulting in effects on the combustion process. Among these fuels are both liquid and gaseous fuels [1, 2].

Rapeseed oil is one of the most widely studied first-generation biofuels. It is produced from rapeseed and has a high viscosity and higher density than diesel. Rapeseed oil is a renewable source of energy, which makes it attractive from a sustainable development perspective. However, its physicochemical properties make a number of modifications necessary in the context of use in diesel engines to ensure combustion stability and energy efficiency. The use of vegetable oil to power an internal combustion engine closes the carbon dioxide cycle in the atmosphere. Oil plants absorb carbon dioxide in the process of photosynthesis, and the internal combustion engine emits it by burning vegetable oils. Of course, we can only talk about the above process theoretically by considering it on a global scale [3].

DOI: 10.1201/9781003466291-9

In order to improve the performance properties of rapeseed oil, researchers have experimented with mixing it with various substances—diesel, ethers, alcohols [4, 5]. The author of the book and his research team decided to add a nonreactive solvent—n-hexane [6–9]. The addition of n-hexane lowers the viscosity of rapeseed oil, which allows it to be better atomized in the injection process and improves the combustion process. The advantage is that refined rapeseed oil can be used without the need for esterification, which simplifies the fuel production process and can reduce costs. In the author's opinion, it is necessary to look for such methods of producing alternative fuels that require a low energy input. This is because it directly translates into expected environmental benefits. An example of oilseed fuel is hydrogenated vegetable oil (HVO), which is basically a synthetic fuel that is produced by hydrogenation processes of vegetable oils. HVO is characterized by its low-carbon and diesel-like physical and chemical properties. This makes HVO an attractive alternative to diesel fuel. Nonetheless, its production is complex and expensive, which poses significant challenges to its widespread use.

Hydrogen is considered one of the cleanest fuels, as no carbon dioxide is produced during its combustion. Hydrogen can be used in diesel engines in dual-fuel systems, where it is burned in the presence of an initiating dose of diesel fuel. This configuration allows a significant reduction in CO_2 emissions, although it requires a precise control of the injection process to avoid problems with premature ignition and engine instability [10, 11].

LPG, or liquefied propane-butane gas, is a popular gaseous fuel that has a high heating value and good miscibility with air. With these properties, LPG promotes efficient combustion and lower emissions of pollutants, including particulates and NOx. Smoke opacity is also reduced. However, it should be remembered that the expected environmental effects depend on the dual-fuel technology used and especially on the control of the gaseous fuel dose and the initiating dose of diesel fuel.

Biomethane, like natural gas, can be used in the form of compressed natural gas (CNG) or liquefied natural gas (LNG). Biomethane production involves the purification of biogas, making it a sustainable energy source. Biomethane and natural gas can be used as fuels for diesel engines with a diesel-initiating dose. The problems with the use of biomethane are similar to those with LPG. An additional problem is its variable quality, depending on the properties of the input feedstock.

In order to accurately assess the effects of various alternative fuels on the operation of diesel engines, a series of experiments were conducted to study their combustion process, energy efficiency, and emissions. These tests were conducted on a specially designed test rig [12], which makes it possible to test various fuels under conditions similar to actual operating conditions. The test stand consisted of several key elements that allowed a comprehensive analysis of the behavior of a diesel engine when running on alternative fuels:

1. The vehicle used for the tests was equipped with a standard diesel engine that was adapted to run on various liquid and gaseous fuels. The engine of this vehicle had a modified power system and an additional gas fuel injection control system, which made it adaptable to different fuel types.
2. The chassis dynamometer allowed the simulation of real driving conditions, including acceleration, braking, and engine load changes. This was a key part of the experiments, as it made it possible to study engine performance under dynamic conditions. Portions of the WLTP test were simulated.
3. The AVL system was used to index the engine, record fuel injection waveforms, and record the engine's crankshaft speed.
4. The MACHA system made it possible to measure the concentration of harmful exhaust components such as CO_2, NOx, HC (hydrocarbon), and smoke opacity. This made it possible to compare emissions for different fuels and assess their environmental impact. Emissions of toxic exhaust components were measured upstream of the exhaust aftertreatment system.
5. The BOSCH system monitored engine diagnostic parameters available from the EOBD system.

The tests were conducted under two main engine operating conditions: static conditions and dynamic conditions. The tests conducted under static and dynamic conditions were aimed at studying various aspects of engine operation when running on alternative fuels. The engine ran on various alternative fuels, and parameters such as power, torque, and exhaust emissions were monitored. In addition, the pressure inside the combustion chamber was analyzed and, based on this, selected thermodynamic parameters were calculated. Dynamic tests were crucial for evaluating engine behavior and exhaust emissions under conditions that better reflect actual vehicle use. During these tests, various driving scenarios, such as acceleration, braking and load changes, were simulated, making it possible to study how different alternative fuels affect engine performance under dynamic conditions. Testing under static and dynamic conditions was conducted to replicate selected portions of the WLTP test.

The results of the experiments provide valuable information on the potential and challenges of using alternative fuels in diesel engines. Analysis of combustion efficiencies of various alternative fuels provides important information on their potential use in diesel engines. In tests using rapeseed oil and its blends with n-hexane, it was observed that the higher viscosity of the fuel increased flow resistance in the injection system, which could affect the performance of the high pump. The use of n-hexane reduced the viscosity of the mixture, which improved the atomization of the fuel and the stability of the combustion process. Test results showed that blends of rapeseed oil with n-hexane could deliver power and torque comparable to conventional diesel, while reducing CO_2 and NO_x emissions. In dynamic tests, blends of rapeseed oil with n-hexane showed satisfactory combustion performance, even during rapid changes in engine speed and load.

The use of hydrogen, due to its unique physicochemical properties, required a sophisticated control system to maintain proper and safe combustion even under dynamic engine operating conditions. One of the main objectives of the tests was to compare exhaust emissions for various alternative fuels. Test results showed that the alternative fuels tested showed potential for CO_2 reduction. This was particularly true for the variants using LPG, hydrogen, and the addition of n-hexane to rapeseed oil. In particular, hydrogen, which contains no carbon, could contribute to global climate goals if properly implemented to power the engine in a dual-fuel system. While the use of LPG led to lower NOx emissions compared to conventional diesel, hydrogen showed a tendency to increase these emissions when burned simultaneously with diesel under higher load conditions. Blends of rapeseed oil with n-hexane and gaseous fuels such as hydrogen had low smoke levels, similar to diesel.

Another important aspect of the research was to assess the compatibility of alternative fuels with diesel engines. The use of some gaseous fuels, such as hydrogen, may require modifications to the design of the diesel engine, including adaptations to the fuel supply system. However, many liquid fuels, such as rapeseed oil blends, can be used without major design changes, making them an attractive solution for retrofitting the existing vehicles. The research further underscores the need to develop the infrastructure for the safe storage and transportation of alternative fuels, especially gaseous fuels such as hydrogen and LPG. These fuels require special tanks and systems to ensure safe operation.

In the course of research on alternative fuels in diesel engines, special attention has been paid to fuel injection technology, which plays a key role in combustion efficiency and emissions. Traditional diesel engines were designed to burn diesel fuel, which has specific physical and chemical properties. The introduction of alternative fuels, such as rapeseed oil, LPG, and hydrogen, required adjustments to injection systems to ensure optimal combustion conditions. Fuel injection in diesel engines is a key process that determines combustion efficiency, emissions, and engine performance. In a traditional diesel engine, diesel fuel is injected under high pressure directly into the combustion chamber, where it is ignited. The use of alternative fuels has required modification of this process to adapt the fuel injection to its physical and chemical properties.

Rapeseed oil, due to its higher viscosity, posed challenges in the context of traditional injection systems [13]. The high viscosity of the fuel increased flow resistance in the injection system, which could lead to insufficient atomization of the fuel, resulting in incomplete combustion and higher emissions. The introduction of n-hexane as an additive to rapeseed oil significantly improved the preparation of the combustible mixture, enabling better mixing with air and more efficient combustion. In the future, additionally, the injection characteristics of mixtures of rapeseed oil with n-hexane, including injection duration and injection pressure, should be adjusted to optimize the combustion process under dynamic conditions. The function of the objective of this optimization should be the obtained energy performance of the engine and the emission of toxic components of the exhaust gas.

As a gaseous fuel, hydrogen differs significantly from diesel fuel in its physical and chemical properties. In dual-fuel systems, where hydrogen is injected along with diesel fuel, a precise control of the fuel dose was crucial. Due to hydrogen's significant rate of combustion and its propensity for premature ignition, the injection system had to be designed to ensure the safety of the combustion process. The study showed that an appropriate use of advanced control algorithms made it possible to achieve stable combustion while reducing CO_2 emissions. However, increased NO_x emissions during hydrogen-diesel combustion indicate the need for further development of injection technology and exhaust aftertreatment systems. Very promising results were obtained with the simultaneous combustion of rapeseed oil with hydrogen and rapeseed oil with n-hexane and hydrogen. Significantly better energy and emission rates were obtained than for the combustion of only rapeseed oil. This indicates new possibilities for using rapeseed oil as a fuel.

One of the key challenges in implementing alternative fuels is the development of adequate infrastructure for their storage and distribution. Alternative fuels, due to their unique properties, require special technical solutions to ensure their safe and efficient use. Hydrogen is a fuel that requires special storage conditions due to its chemical and physical properties. As the lightest element, hydrogen is difficult to store because it requires high pressure or low temperature to keep it in a liquid or compressed state. The use of hydrogen as a fuel in diesel engines necessitates the development of advanced tanks that can withstand pressures of 700 bar or more. These tanks must be made of high-strength materials, such as carbon composites, which at the same time allow lightweight construction, which is key to minimizing additional vehicle weight. In addition, hydrogen storage requires special safety valves to prevent excessive pressure buildup in the tanks and monitoring systems to check for leaks and detect possible leaks. In the case of hydrogen transportation, it is also necessary to develop a network of refueling stations equipped with the appropriate technologies to compress and distribute hydrogen.

LPG, as a liquefied gas, is a more widely used alternative fuel than hydrogen, but it requires specialized infrastructure for storage and distribution. LPG is stored in special pressurized tanks to keep it in a liquid state. These tanks must be equipped with safety valves and monitoring systems to prevent the risk of explosion in case of leaks. LPG-refueling stations must be equipped with appropriate dispensers that are capable of delivering fuel at the right pressure to the vehicle's tank. Another important aspect is the development of a distribution network that will ensure the availability of LPG throughout the country, which is key to the spread of this fuel in road transport.

Biomethane, like natural gas, can be stored in compressed (CNG) or liquefied (LNG) form. Storing biomethane in compressed form requires special tanks that can withstand pressures of 200–250 bar. These tanks must be made of durable materials that also reduce the overall weight of the vehicle. Storing biomethane in a liquefied form requires keeping it at a low temperature, which requires advanced cryogenic technology. LNG tanks must be well insulated to prevent the evaporation of the fuel, and refueling stations must be equipped with appropriate technologies for liquefying and distributing biomethane.

Against the backdrop of global trends in the automotive and energy sectors, alternative fuels are a key component of sustainable development strategies. Increasing demand for clean energy and tightening emissions regulations are prompting both vehicle manufacturers and governments to invest in the development of alternative fuel technologies.

As a zero-CO_2 fuel, hydrogen is seen as one of the most promising solutions for decarbonizing transportation. Many countries, including Japan, Germany, and South Korea, have invested heavily in the development of hydrogen infrastructure, including the construction of hydrogen-refueling stations and the development of hydrogen storage and transportation technologies. One of the key challenges associated with hydrogen is its production. Currently, most hydrogen is produced from natural gas by steam reforming, which is associated with CO_2 emissions. However, the development of electrolysis technology, which makes it possible to produce hydrogen from water using renewable energy, opens up new opportunities in the context of producing "green" hydrogen that does not generate greenhouse gas emissions. In the automotive sector, hydrogen can be used in both fuel cells and internal combustion engines in dual-fuel systems. However, the introduction of hydrogen into the transportation sector requires significant investment in infrastructure and further research into optimizing injection and combustion technologies for this gas. In the long term, hydrogen could play a key role in the decarbonization of trucking, shipping, and aviation.

LPG, due to its availability and developed infrastructure, is often seen as an interim fuel on the road to full transportation decarbonization. LPG offers the benefits of lower CO_2 emissions compared to diesel, as well as lower NOx (provided the increase in combustion temperature is limited) and particulate matter. These properties make LPG an attractive option for countries seeking to rapidly reduce emissions from the transportation sector without requiring significant investment in new technology. Despite its numerous advantages, LPG is not a completely emission-neutral fuel, as its combustion still generates some CO_2. Therefore, LPG can act as a transitional fuel that will gradually be replaced by more advanced and cleaner technologies such as hydrogen or electricity from renewable energy sources.

Biomethane, due to its renewable origin, is seen as a fuel with great potential in the context of sustainable development. The production of biomethane from organic waste, such as agricultural or municipal waste, contributes to the reduction of greenhouse gas emissions and a closed-loop economy. Biomethane can be used both in internal combustion engines and in cogeneration systems, allowing its wide application in transportation, various industries, and power generation. One of the main challenges associated with biomethane is the development of infrastructure for its production, storage, and distribution, which requires significant investment. Improving the design of internal combustion engines in cogeneration systems is also an important issue. It is necessary to work on improving their performance characteristics in the context of feeding these engines with biogas with variable physicochemical properties.

Global emissions regulations and climate change are forcing governments and companies to take action to reduce greenhouse gas emissions. In this context, alternative fuels play a key role in achieving sustainability goals.

The European Union has adopted ambitious climate goals to achieve climate neutrality by 2050. As part of this strategy, the EU is promoting the development of alternative fuels such as hydrogen, biomethane, and renewable electricity. Financial support programs, such as the Innovation Fund, and regulations, such as the Directive on the Promotion of the Use of Energy from Renewable Sources (RED II), are stimulating investment in low-carbon technologies and infrastructure development.

In Asian countries such as Japan and South Korea, hydrogen is playing an increasingly important role in energy strategies. Japan was the first country in the world to announce a plan to create a hydrogen society, in which hydrogen will be the main source of energy for transportation, industries, and households. Investment in hydrogen technology research and infrastructure development, including hydrogen-refueling stations, are key elements of these strategies.

In the United States, energy policy emphasizes the development of renewable energy technologies, including biofuels and hydrogen. Federal programs, such as the Advanced Research Programs Agency-Energy (ARPA-E), support innovation in alternative fuels, which contributes to the development of new technologies and their commercialization. In addition, initiatives such as the California Hydrogen Highway aim to develop hydrogen infrastructure at the state level, an important step toward popularizing hydrogen as the fuel of the future. In the United States, as in Europe and Asia, the development of hydrogen and biomethane technologies is seen as a key element in achieving climate goals.

In the long term, alternative fuels have the potential to replace traditional fossil fuels in diesel engines, especially in sectors such as trucking and marine and industrial transportation. However, achieving this goal will require further investment in research, technological development, and infrastructure. As technology powered by alternative fuels continues to develop, smart energy grids will play an increasingly important role in optimizing energy consumption and integrating various energy sources, including renewable and alternative fuels. The development of such grids will allow for more efficient management of energy resources, which will be crucial in the context of global efforts to reduce emissions. Smart energy grids, also known as smart grids, can integrate a variety of energy sources such as wind farms, solar panels, and hydrogen production from electrolysis, enabling a flexible management of energy supply and demand. In the context of transportation, such grids can also support the development of electric-vehicle-charging infrastructure and hydrogen-refueling stations, ensuring the stability and efficiency of the entire energy system.

Increasing public awareness of the benefits of alternative fuels will be key to their widespread adoption. Educational campaigns, support for research, and initiatives to increase the availability of information on alternative fuels will help accelerate the transition to sustainable energy sources. The public plays a key role in the energy transition, and their understanding and acceptance of new technologies is essential to their success. Therefore, it is important that both governments and NGOs, as well as industry, actively engage in public education about alternative fuels, their advantages, challenges, and environmental impact. Education should cover all aspects of alternative fuels, including their production, storage, distribution, and use. It is also important to provide information on technological advances that can make these fuels more accessible and economically viable.

Global climate and sustainability challenges require international cooperation in research, technology development, and the creation of standards and regulations. International partnerships, such as the International Renewable Energy Agency (IRENA) and the International Energy Agency (IEA), play a key role in stimulating innovation and knowledge sharing in the field of alternative fuels. International cooperation enables the exchange of best practices, technology transfer, and joint financing of R&D projects, which accelerates the process of implementing innovative solutions around the world. Such partnerships also make it possible to harmonize regulations and standards, facilitating the global integration of new technologies and supporting the development of the global alternative fuels market. One example of successful international cooperation is the "Mission Innovation" initiative, which involves countries from around the world aiming to double the investment in research and development of clean energy technologies over the next decade. The initiative focuses on supporting research into new fuels, energy storage technologies, and energy systems that can make a significant contribution to reducing global emissions.

Although alternative fuels offer great potential for reducing emissions and improving energy efficiency, their widespread use faces numerous technological, economic, and social challenges. The development and deployment of alternative fuels require an integrated approach that considers everything from basic research to commercialization and infrastructure development. One of the main technological challenges is optimizing the combustion of alternative fuels in diesel engines. These fuels, depending on their physicochemical properties, may require modifications to injection and ignition systems, as well as exhaust aftertreatment systems. Hydrogen, for example, due to its propensity for premature ignition, requires a precise control of the injection and combustion processes, which poses significant engineering challenges. In addition, the development of advanced storage technologies for fuels such as hydrogen and biomethane is essential to ensure their safe and efficient use. New materials and technologies need to be developed to enable the construction of lightweight, durable tanks and the development of fueling and distribution infrastructure.

Another challenge is the development of flue gas cleaning technologies, which may be particularly needed when using hydrogen, which contributes to NOx emissions. New technologies can play a key role in minimizing emissions and meeting increasingly stringent environmental standards.

The costs of producing, storing, and distributing alternative fuels are currently higher than for traditional fossil fuels. The high costs of HVO production and advanced hydrogen storage infrastructure are significant barriers to their widespread deployment. In order for alternative fuels to become competitive, it is necessary to continue to invest in cost reduction research and to develop business models that address both economic and environmental considerations. Government subsidies, tax breaks, and other forms of financial support can play a key role in reducing costs and stimulating investments in alternative fuel development. Many countries are already introducing support programs for the development of clean energy technologies, helping to accelerate their large-scale deployment. One example of such measures is the financial support for the development of hydrogen infrastructure in Europe, where the European Commission has launched programs to finance the construction of hydrogen-refueling stations and research into new technologies for producing the fuel.

The social challenges of implementing alternative fuels include issues of public acceptance, education, and changes in the labor market. The energy transition requires not only investment in new technologies but also a change in public thinking and attitudes toward new energy sources. Public acceptance of alternative fuels is crucial to their success. Educational campaigns that raise awareness of the benefits of these fuels and their impact on the environment can contribute to greater public acceptance and accelerate their adoption. In addition, the development of new alternative fuel technologies can lead to changes in the labor market, including the creation of new jobs in sectors related to the production, distribution, and operation of infrastructure for alternative fuels. For this reason, it is important for governments and companies to invest in training programs that will prepare workers for jobs in new, rapidly developing sectors.

Future research on alternative fuels in diesel engines will focus on several key areas. Further work is needed on optimizing the combustion process, including the development of advanced injection and combustion systems that are tailored to the specific properties of alternative fuels. Research into new fuel storage materials that will allow the construction of lighter, more durable, and safer tanks will be important. In particular, the development of composite materials and cryogenic technologies will be crucial for hydrogen and biomethane storage. Research is needed on new exhaust aftertreatment technologies that can meet the challenges of burning hydrogen and other alternative fuels. Development of catalytic converters, particulate filters, and NOx reduction systems will be key to meeting future emission standards.

Research on alternative fuels in diesel engines is providing valuable information on the opportunities and limitations associated with their use in transportation and industries. Fuels such as hydrogen, LPG, biomethane, and blends of rapeseed oil with n-hexane have great potential for reducing emissions and improving energy efficiency. However, their widespread use will require overcoming numerous technological, economic, and social challenges. In the long term, alternative fuels can play a key role in the decarbonization of transportation, but their implementation will require an integrated approach that takes into account the development of technology, the construction of appropriate infrastructure, and the education of the public. International cooperation, government support, and further research will be key to achieving these goals and creating a sustainable energy system based on alternative fuels. As these technologies develop, we can expect to see more and more integration of alternative fuels into smart grids and their growing share in the global energy mix.

Education, innovation, and international cooperation will be key enablers for realizing the vision of a world in which transportation is based on sustainable, clean energy sources, helping to protect the environment and improve the quality of life for future generations.

REFERENCES

[1] J. Martins and F. P. Brito, "Alternative fuels for internal combustion engines," Proceedings of the Combustion Institute, vol. 36, 2020, doi: 10.3390/en13164086.
[2] J. C. W. Lan, A. Tsui, H. S. Wang, and S. S., Wu, "A review of biodiesel as renewable energy," Biochemical Engineering, pp. 1–40, 2009.
[3] C. L. Peterson and T. Hustrulid, "Carbon cycle for canola oil biodiesel fuels," Biomass Bioenergy, vol. 14, no. 2, 1998, doi: 10.1016/S0961-9534(97)10028-9.
[4] R. Smigins and A. Zakis, "Impact of diethyl ether/canola oil blends on performance and emissions of a light-duty diesel vehicle," Energies (Basel), vol. 13, no. 15, 2020, doi: 10.3390/en13153788.
[5] İ. Sezer, "A review study on using diethyl ether in diesel engines: Effects on fuel properties, injection, and combustion characteristics," Energy and Environment, vol. 31, no. 2, 2020, doi: 10.1177/0958305X19856751.
[6] R. Longwic and J. Kowalczyk, "The influence of the dynamic angle of fuel pumping start on selected parameters of the combustion process in diesel engine powered by mixtures of canola oil with n-hexane," IOP Conference Series: Materials Science and Engineering, vol. 421, 2018, doi: 10.1088/1757-899X/421/4/042049.
[7] R. Longwic and P. Sander, "The course of combustion process under real conditions of work of a traction diesel engine supplied by mixtures of canola oil containing n-hexane," IOP Conference Series: Materials Science and Engineering, vol. 421, 2018, doi: 10.1088/1757-899X/421/4/042050.
[8] K. Górski, P. Sander, and R. Longwic, "The assessment of ecological parameters of diesel engine supplied with mixtures of canola oil with n-hexane," IOP Conference Series: Materials Science and Engineering, vol. 421, 2018, doi: 10.1088/1757-899X/421/4/042025.
[9] R. Longwic, P. Sander, and D. Tatarynow, "Ecological aspects of using mixtures of canola oil with n-hexane in diesel engine," Combustion Engines, vol. 190, no. 3, 2022, doi: 10.19206/CE-143245.
[10] S. H. Hosseini, A. Tsolakis, A. Alagumalai, O. Mahian, S. S. Lam, J. Pan, W. Peng, M. Tabatabaei, and M. Aghbashlo, "Use of hydrogen in dual-fuel diesel engines," Progress in Energy and Combustion Science, vol. 98, p. 101100, 2023, doi: 10.1016/j.pecs.2023.101100.
[11] V. N. Nguyen, S. K. Nayak, H. S. Le, J. Kowalski, B. Deepanraj, X. Q. Duong, T. H. Truong, D. N. Cao, and P. Q. P. Nguyen, "Performance and emission characteristics of diesel engines running on gaseous fuels in dual-fuel mode," International Journal of Hydrogen Energy, vol. 49, pp. 868–909, 2024, doi: 10.1016/j.ijhydene.2023.09.130.

[12] D. Tatarynov, R. Longwic, P. Sander, Ł. Zieliński, M. Trojgo, W. Lotko, and P. Lonkwic, "Test stand for a motor vehicle powered by different fuels," Applied Sciences (Switzerland), vol. 12, no. 20, 2022, doi: 10.3390/app122010683.

[13] A. Zdziennicka, K. Szymczyk, B. Jańczuk, R. Longwic, and P. Sander, "Adhesion of canola and diesel oils to some parts of diesel engine in the light of surface tension components and parameters of these substrates," International Journal of Adhesion and Adhesives, vol. 60, 2015, doi: 10.1016/j.ijadhadh.2015.03.001.

Index

For Product Safety Concerns and Information please contact our EU representative GPSR@taylorandfrancis.com
Taylor & Francis Verlag GmbH, Kaufingerstraße 24, 80331 München, Germany

www.ingramcontent.com/pod-product-compliance
Lightning Source LLC
LaVergne TN
LVHW020651100826
845148LV00012B/2439

* 9 7 8 1 0 3 2 7 3 8 5 6 7 *